THE YOUNG OXFORD BOOK OF

THE *human* BEING

THE YOUNG OXFORD BOOK OF

THE *human* BEING

David Glover

OXFORD UNIVERSITY PRESS

CONTENTS

Oxford University Press

Oxford   New York
Athens Auckland Bangkok Bogotá Bombay
Buenos Aires Calcutta Cape Town Dar es Salaam Delhi
Florence Hong Kong Istanbul Karachi
Kuala Lumpur Madras Madrid Melbourne
Mexico City Nairobi Paris Singapore
Taipei Tokyo Toronto Warsaw

and associated companies in
Berlin Ibadan

Copyright © David M. Glover 1997
Published by Oxford University Press, Inc.
198 Madison Avenue, New York, New York 10016

Originally published by Oxford University Press UK in 1996

Oxford is a registered trademark of Oxford University Press

Library of Congress Cataloging-in-Publication Data

Glover, David M.
The Young Oxford Book of the Human Being / David Glover.
p. cm.
Includes index.
1. Human physiology—Juvenile literature. 2. Human
evolution—Juvenile literature. 3. Sociology—Juvenile
literature. 4. Psychology—Juvenile literature.
[1. Human physiology. 2. Body, Human. 3. Evolution.
4. Sociology. 5. Psychology.]
I. Title.
QP37.G56 1997
612.8—dc21   97-11922
                                  CIP
                                  AC

ISBN 0-19-521375-0 (trade ed.)
ISBN 0-19-521374-2 (lib. ed.)

9 8 7 6 5 4 3 2 1

Printed in Italy by G. Canale & C. S.p.A. - Borgaro T.se - TURIN

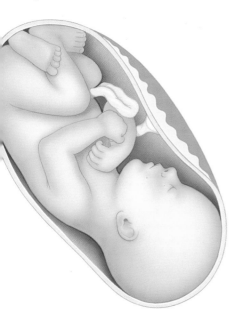

3
MIND

4
LIVING TOGETHER

INTRODUCTION

This is a book about us – human beings. What are we? Where do we come from? How do we fit into the world? For most of human history people have tried to make sense of their lives through mythical stories of the past, beliefs about nature or faith in a particular religion. Today, however, we can seek answers to many questions about ourselves in another way – through science.

Science can tell us an amazing amount about our origins, our bodies, our minds and how we live together. Within the space of just a few generations we have begun to trace the human family tree from our earliest ancestors to the present day. We have unravelled the genetic code that makes each of us unique. We have investigated the human body to find out how it is put together, how its parts function and why we sometimes get ill. Now scientists are tackling one of the greatest challenges of all – understanding the human mind. Some scientists believe that one day we will have scientific explanations for our thoughts and feelings, but others are not so sure.

Science is an incredibly powerful tool for studying ourselves and our place in the universe. But there will always be some questions that science cannot answer. Questions such as 'Why do I exist?' or 'What does my life mean?' are beyond scientific understanding. The solutions to problems like these, if they are to be found at all, are to be discovered within us, and each one of us will have our own answers.

David Glover

ORIGINS

Where did human beings come from?
How long have we existed? Do we all share the
same origin? How are we related to other living
things? Now that we have evolved to be able to
explore and understand our world, science is
providing some of the answers to these
fascinating questions.

THE TREE OF LIFE

In the past, human beings used to think of themselves as separate from the rest of Nature, as different and special. But the theory of evolution and the discovery of the genetic code have shown that we share our origins and life processes with all living things on the Earth. We are a recent, and very successful, branch of the tree of life.

What are human beings? In many ways the human being is a very ordinary animal. Physically, we are far from being the strongest, fastest or most agile animal on the planet. In the animal Olympics we would not win any medals for speed or strength. In the history of life on Earth we are mere newcomers. The dinosaurs were the dominant land animals for nearly 200 million years. Sharks have prowled the oceans for more than 400 million years. Yet human beings have been around for just 2 to 3 million years, with modern humans like ourselves appearing as little as 100,000 years ago. We do not even have a particularly long life span. Parrots, elephants and whales can live as long as, or longer than, us. The Galapagos tortoise, for example, can live for 150 years.

So why do we think we are so special? Numbers for one thing. There are currently more than 5000 million human beings on the Earth, many more than any other kind of medium- to large-sized animal. (The human population is, however, small in comparison to that of insects and other small living things – there may be 2 million termites in a single mound!) But, without doubt, the key to the success of humankind is our large, complex brain and the capacities it gives us for communication, organization, creativity and invention. No other creature has the range of abilities of the human being, from using fire, making tools, constructing shelters and growing food, to clothing its body, building vehicles and adapting the environment to its needs. Our unique mental skills have enabled us to spread throughout the world, becoming the dominant species on the planet in just a few thousand generations.

The living planet

We cannot hope to understand the human being in isolation. We are part of the living world and we need to answer some questions about all living things before we can understand ourselves.

▷ The human urge to explore the unknown has taken us to every part of the Earth, from the desert sands to the polar icecaps. Now we are using our intelligence and inventiveness to venture into space.

▷ Through the sounds of music human beings can express moods and emotions. Our capacity to communicate feelings, ideas and discoveries through speech, literature and art is far greater than that of any other species. Our culture, the knowledge which we share and pass on down the generations, is an important part of what makes us human.

What exactly is life? What separates something that is living, like you, from a lifeless object such as a rock or a bicycle? Life is difficult to define precisely, but some of its features include the following.

Living things are called 'organisms'. They have bodies that organize them-selves into separate parts to do special jobs – parts for moving, feeding, reproducing and so on. You could argue that a car is organized in this way, but the difference is that someone must design the car and put its parts together. The parts of a living organism assemble themselves – it is a 'self-organizing' structure.

Living things need energy and raw materials to grow and to maintain their bodies. They take in energy from sunlight or food, and dispose of waste materials.

Living things reproduce. They make copies of themselves, which in turn feed, grow and reproduce to make further generations. You are one generation of human being, your parents were the previous generation that produced you, and your children will belong to the next generation.

Nature's kingdoms

The sheer variety of life on the planet is almost overwhelming. Nearly 2 million different species have been identified and many more have yet to be discovered by science. A species is a very exclusive club. To be a member of a species you must be able, in principle at least, to reproduce by mating with another member of the species. The members of one species cannot reproduce with members of another species. For example, a horse cannot reproduce with a rabbit, and neither can reproduce with an oak tree. To make sense of the great diversity of life scientists search for links between species and group living things according to their similarities.

All living things on Earth belong to one of just five kingdoms – Monera, protists, fungi, plants and animals.

Monera and protists are microscopic life forms. You cannot see them with the naked eye, but they are the most numerous inhabitants of the planet, living in the soil, in water and in the living and dead bodies of other organisms, including our own. The bodies of these microbes consist of a single cell. A cell is the smallest unit of life, like a tiny packet of living jelly, but inside it has all the machinery needed to take in and process energy, excrete waste and to reproduce itself by dividing in two.

Bacteria are members of the Monera kingdom. A typical bacterium is just a few thousandths of a millimetre across – several hundred bacteria would fit across one full stop on this page. Some bacteria are harmful to human beings, causing diseases such as cholera, food poisoning and tetanus, but others play an important role in our digestion and in the breakdown of waste materials.

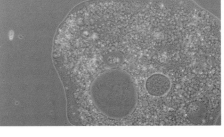

△ The amoeba is a protist that lives in water. This amoeba can change its shape to move and to engulf its food.

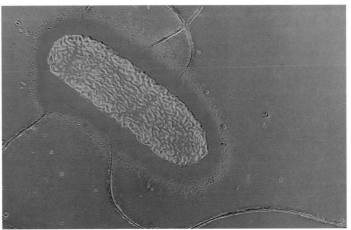

◁ Micro-organisms, such as these salmonella bacteria which can contaminate food, are by far the most numerous living things on the planet.

△ The plant kingdom contains some of the most beautiful of all living things, such as these orchids.

◁ This spectacular growth is the fruiting part of a fungus, which is feeding on a forest tree.

living things – mammals, such as the great whales, elephants, chimpanzees and ourselves.

Origin of species

All human beings are members of the same species (*Homo sapiens sapiens*). There are many different species. How did this tremendous variety come about?

Until the middle of the last century a popular view was that all species on Earth had always been just as they are. Nature was unchanged since the Earth had been created, perhaps 6000 years ago. Each species had a place in the scheme of life. Lowly life forms crawled on the ground and slithered in the mud. Humans, superior beings, were placed above the rest of the natural world. But, with the emergence of science and its methods of careful observation and experiment, this view of Nature was questioned. A growing body

The third kingdom is that of fungi. Some fungi, for example yeasts, are single-celled, but most are 'multicellular' organisms. This means that they are built from many cells that work together as a single living thing. We are familiar with the mushrooms and toadstools that grow out of the soil, but these are only a part of a fungus structure. The main part of a fungus is a network of fine threads that spreads through the soil. Fungi feed on dead leaves, fallen trees and the rotting remains of animals, absorbing chemicals from their bodies as they break down.

Plants are Nature's food factories. The members of the plant kingdom are the only living things that make their own food from non-living raw materials. Plant cells contain a chemical called chlorophyll that absorbs the energy of sunlight. The plant uses this energy to combine carbon dioxide from the air with water, to make sugars. This process is called 'photosynthesis', and it is the source of the food that fuels all life on Earth.

The fifth kingdom, and the kingdom to which human beings belong, is the animal kingdom. Animals are multicellular

organisms that obtain their food by eating plants and each other. They are Nature's consumers. Animals are much more mobile than plants, propelling themselves by slithering, creeping, swimming, walking on limbs or even flying with wings. Like plants, animals come in an incredible variety, from simple microscopic worms, to strange transparent sea creatures and the most complex of all

▷ Dolphins have highly developed brains, built from nerve cells like those of human beings. All but the simplest members of the animal kingdom have a nervous system, to coordinate body movements and to process information from the environment.

NATURE'S WEB

A living thing cannot survive on its own. Some science-fiction stories describe metal-clad planets inhabited only by futuristic humans, all other life forms having been eliminated. But whoever imagined these future worlds did not understand how life works. Life on Earth is maintained by a complex web of links between the species of all five natural kingdoms. For example, in the rainforest plants are the 'primary producers' and animals are the 'consumers', eating the plants and each other, and breathing oxygen to supply the energy they need. But animals do not just take from the system. Their activities carry pollen between flowers and disperse seeds to help maintain the plants' cycle of growth and reproduction. Fungi and microbes are 'decomposers' on the forest floor and in the soil. They speed up the process of natural decay, returning nutrients from animal droppings, fallen leaves and the bodies of the dead to the soil, making them available again for the plants.

Humans beings are just as much a part of Nature's web as any other living thing. If we cut our links to the other species on Earth, by destroying natural environments or by moving to lifeless worlds in space, then we cut off our food supply and our source of fresh air. A future world colonized by humans will be a green and living one like Earth, not a sterile landscape of concrete and steel.

△ All life on Earth depends on the energy of sunlight trapped by the green leaves of plants. As members of the animal kingdom we humans obtain our energy secondhand by eating plants and other animals.

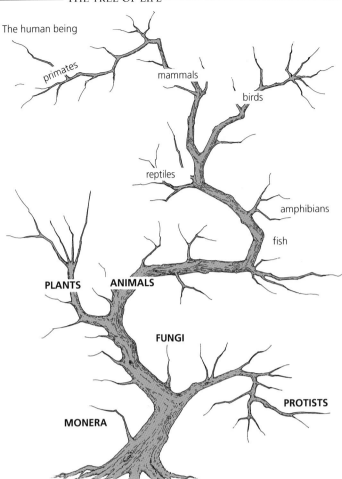

SELF-COPYING MOLECULES 4000 billion years BC

of evidence showed that the Earth was really much, much older than it had been thought. It began to emerge that far from being constant, living species changed, or evolved, through time.

In 1859 the English scientist Charles Darwin published one of the most important of all science books, *On the Origin of the Species by Means of Natural Selection*. In this book Darwin describes his theory of evolution.

Darwin's theory of evolution is that, far from being created just as they are, each living species has developed from earlier forms of life by a process of change. The key to the process is that offspring (a new generation of living things) are not identical to their parents – if they were there would be no change and no evolution. Life has evolved over many millions of years. Starting from very simple life forms, slight changes at each generation have gradually produced new species. One species branches into two, each of which may branch again into more new species. Some species are more complex than their

△ This 'tree of life' gives an impression of the evolution of the millions of species, some living, but most extinct, that natural selection has produced. The human species is a tiny bud on a branch near the top, showing just how recently we have appeared in the story of life on Earth.

1. Geospiza magnirostris
3. Geospiza parvula.

2. Geospiza fortis.
4. Certhidea oliva3ea.

Life's thread

Darwin's great theory explained how species evolved as a result of natural selection. But, nearly 150 years ago, he had little idea of exactly how the features of parents are inherited by their children, nor what produces the variations and mutations that, through natural selection, eventually lead to new species.

Then, in 1953 in Cambridge, Francis Crick and James Watson made what many scientists regard as the single most important scientific discovery of the 20th century. Crick and Watson worked out the structure of the DNA molecule. DNA (deoxyribonucleic acid) is a microscopic thread that exists inside the cells of every living thing. The importance of the discovery was that it showed how DNA acts both as a blueprint for life, by storing the plans that build living bodies, and as a messenger to carry these plans from generation to generation.

Imagine trying to write down the instructions needed to build your own body. You might start with a description of your external appearance – two arms, two legs, brown eyes, black hair. But that would be only a minute fraction of the information required. The details of your development and growth would have to be planned, the many thousands of chemicals from which your body is built listed, the jobs performed by your internal organs specified and the special features that make you different from every other living

predecessors, some can live in a different habitat and some, such as the various species of crocodile, survive almost unchanged for millions of years. Some species become extinct, perhaps by evolving into new species that replace the old one, or by dying out altogether, as every single species of dinosaur did.

Natural selection

One of the puzzles that the theory of evolution must explain is the huge variety of life. Darwin recognized that living things can vary from generation to generation. For example, you may grow up to be either taller or shorter than your parents. He also knew that occasionally a completely new feature appears in species, such as a flower colour not present in previous generations. This is called a 'mutation'. Plant breeders are constantly on the lookout for mutations to develop new varieties. The breeders select their seeds from the plants with the characteristics they like.

Darwin's theory was that in Nature new species arise by a process he called 'natural selection'. He observed that nearly all living things produce many more offspring than eventually survive to reproduce themselves. Those that do survive

△ In 1835, on the isolated Galapagos Islands, Charles Darwin observed 13 similar species of finch-like bird with different beak shapes. He concluded that their beaks had been adapted to enable them to obtain different foods, and that the different species had all evolved from a common ancestor.

are the strongest, or fittest. If the offspring pass on to their offspring any of the differences they have that make them better able to survive, then these offspring will also be more likely to survive. For example, the young of a particular species of fish may be eaten by another species. The ones that stand the best chance of growing up are large fish that can swim fast, or small fish that can hide in holes in the rocks. If the small fish also tend to have small offspring and the large ones large offspring, then the fish may eventually evolve into two separate species: a species of large fish and a species of small ones.

Is it really possible that, starting from the first simple living things, tiny changes at each generation have eventually produced things as different as a shark and a daisy? Some opponents of evolution do not believe that this could happen by chance. They argue that the variety of life must indicate a guiding force, perhaps a supernatural one, behind evolution.

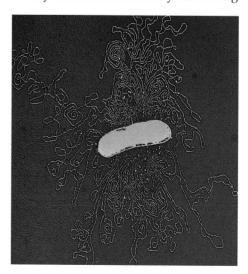

△ The DNA threads which carry the genetic program of this bacterium have exploded from its body. In a living bacterium the DNA is tightly coiled inside.

PLANS FOR LIFE

Life's plans are written along the two spiral strands of the DNA molecule. Each spiral is known as a 'helix'. The two strands of the molecule are held together by pairs of chemical 'hooks' called bases. There are four kinds of these bases. Each base will only hook to one of the other three types, so the bases along one thread must be paired in the correct way with those on the other.

We can understand how DNA works by picturing it as a long list of instructions like a computer program. The bases are the 'alphabet' with which the program is written. Each word of the program consists of just three letters from the alphabet. Strings of these words form instructions used by our body cells to construct the molecules that build our bodies. These instructions are the 'genes' that determine what your body is like.

So just how much information is needed to design a human being? There are more than 3000 million letters of genetic code along the DNA strands that make a human being. This is roughly equal to the number of letters in 10,000 copies of this book. Using modern technology you could, in principle, fit the plan for your body on to five CD-ROMs but, in comparison to DNA, this is still a very cumbersome way to store data.

▽ The two spiral strands of the DNA molecule can separate, or 'unzip'. Each strand can then form a new double helix. In this way DNA copies itself. One double helix becomes two, each carrying identical information.

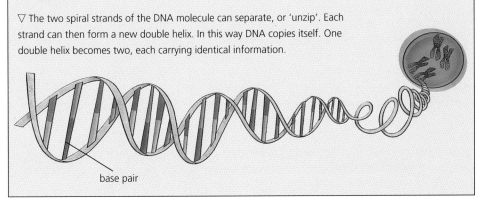

base pair

Never-ending copies

The remarkable double helix structure of DNA allows the molecule to copy itself. Scientists say that it is 'self-replicating'. When a simple organism such as a bacterium divides, its DNA divides and replicates, so that each of the new bacteria contain the same genes as the 'parent'. Occasionally a mistake or mutation is made in the genetic code and a bacterium acquires a characteristic that was not present in the parent. Usually this mutation will be harmful and the bacterium will not survive. But very occasionally the mutation will give the bacterium a survival advantage. In this way a successful new form, or strain, of bacteria

thing described. What is incredible is that all this information is stored on a tiny thread, too small to see with the naked eye. Copies of it are found in nearly every cell in your body.

is started. We can see this happening today when flu viruses mutate and new strains of flu arrive each year.

More complex life forms, such as birds, fish and humans, reproduce sexually. In sexual reproduction segments of the DNA molecules from two different individuals link together, shuffling their genes. Each offspring is a unique individual (except when it is an identical twin) with a mixture of its parents' features. New combinations of characteristics, previously untried by Nature, may give survival advantages, for example in resisting new viruses.

This then is how the evolution of life has taken place. Four billion years ago, chance chemical reactions in the oceans formed a simple DNA molecule capable of self-replication. Once started, replication has continued to this day, with subtle changes at each generation. All living things today contain similar DNA and are therefore almost certainly descended from this earliest life form.

A new understanding

Using sophisticated methods, scientists can now read the genetic progam of a living creature. It is a long and tedious job because there are millions of letters of genetic code to read on a single DNA strand. But deciphering the genes has led to a new understanding of the links between species and the story of evolution. Shared genes suggest shared ancestors; the more genes two species share, the more recently they branched from their common ancestor on the evolutionary tree. We humans, for example, share most of our genes with the chimpanzee. Now, the Human Genome Project is attempting to read the complete genetic program of the human being. This international project, involving scientists from many nations, will both increase our understanding of what makes us human and produce information needed in the fight against disease.

▷ Shared genes prove that every living thing is related. We humans share most of our genes with our nearest relative, the chimpanzee. Both chimps and humans share genes in common with a banana.

ANCESTORS AND RELATIVES

The search for our ancestors is a fascinating detective story. The evidence lies in their fossilized bones and in our genes. The clues are sometimes hard to interpret, but a picture of 7 million years of human evolution is now starting to emerge.

Have you ever tried to draw your family tree? Many people do. But why are we so interested in researching our ancestors and locating our relatives?

Natural curiosity means that we want to know more about ourselves and our roots. Our ancestors' genes have passed through the generations into our bodies, so learning about our ancestors perhaps tells us something about ourselves and our potential. Our living relatives share genes in common with us too, and they act as mirrors in which we see ourselves, our strengths and weaknesses reflected.

Researching the family tree of the whole of humanity gives us similar insights on a grander scale. By learning how we descended from, and how we relate to the rest of Nature we can understand ourselves better. The family tree of the human being is still not understood fully, but the process of discovering it has already removed many myths about the superiority of one human group over another. In fact, the story of evolution shows just how closely all human beings are related and how much we share in common with other species.

Finding our place

To place ourselves in relation to other species we can use a scheme called a taxonomy. This is a system by which living things are divided into a series of smaller and smaller groups according to features they have in common. Members of a group are assumed to share a common ancestor. The largest groups are the five natural kingdoms (see pages 9 to 10).

Within the animal kingdom, humans are vertebrates, tetrapods (we have four limbs) and belong to the 'group' called mammals. We belong to the 'order' of primates, which includes lemurs, lorises, monkeys and apes. Within the order of primates our 'family' is that of the great apes together with the orang-utan, the gorilla and the chimpanzee. The final divisions are the 'genus' and the 'species'. *Homo sapiens sapiens* (the species name for modern humans) is the only living

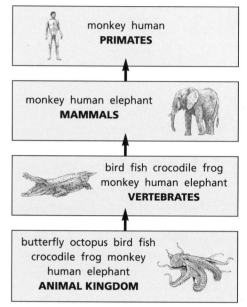

△ Human beings are 'vertebrates', which means that like fish, birds, reptiles and amphibians we have a backbone. We are also 'mammals' (we have hairy bodies and feed our babies with milk) and 'primates' like monkeys and apes.

▽ Are these your ancestors? Fossil skulls like these provide clues in the search for human origins.

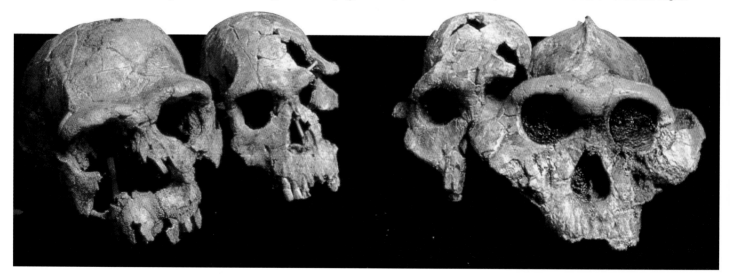

member of the genus *Homo*, but it is not the only human species that has ever existed. Fossil remains of several extinct *Homo* species have been found, some of which may have been our direct ancestors. These extinct species can be placed on the taxonomic chart by comparing their physical features to those of living species and other fossils. But before discussing our extinct relatives, we will look at our living relations, who tell us much about ourselves.

Close relatives

The taxonomic chart for the human being shows that the living species we share most in common with are the other great apes – gorillas, orang-utans and, closest of all, chimpanzees. The more we study chimpanzees, the more like us they appear. There is an obvious physical resemblance and we find their human-like faces very appealing. Like us, they can move their thumbs towards their fingers to grasp and manipulate objects. They are probably the next most intelligent living species after ourselves. They live in social groups called bands. They eat a mixture of plants, occasionally hunt other animals, and can learn to use simple tools to crack nuts and extract termites from mounds.

When Darwin first published the theory of evolution, the idea that human beings and apes were related created much controversy. In a famous incident at a public lecture in 1860, the Bishop of Oxford tried to insult Thomas Huxley, one of Darwin's supporters, by asking if he was related on his grandfather's or his grandmother's side to an ape! Today we could give the truthful answer – both. The theory of evolution makes it clear that we all share common ancestors with the apes.

Just looking like chimpanzees does not prove that we are related. We could have the same shaped face and hands just because they suit the way we live. Sharks and dolphins, for example, are not closely related at all, but they have evolved similarly shaped bodies to swim efficiently. But the chance that our genes (see page 13) are almost identical to those of another species is incredibly small, unless we have inherited our DNA from the same ancestor. There are so many genes along a strand of DNA that the chance of two strands from completely unrelated origins having the

▷ ▽ We share many common features with the other great apes – chimpanzees, gorillas and orang-utans. But these living apes are our cousins, not our ancestors.

Until recently the gorilla (below) was thought to be our closest living relative. Now genetic tests show that humans and chimpanzees (right) have a more recent common ancestor than humans and gorillas, who lived about 7 million years ago.

▽ Our relationship to the orang-utan is more distant than to the other great apes. Our common ancestor with orang-utans lived 12 to 15 million years ago.

same sequence is much more remote than winning the lottery.

Genetic scientists are now able to compare pieces of DNA from different individuals and species, to judge how closely they are related. This method shows that we share 98 per cent of our genes with chimpanzees, confirming that we are close relations. It also proves that, despite the physical differences between individual humans, we all have remarkably similar DNA, with perhaps only one letter in a thousand of the genetic program different between all the people on Earth. This confirms that all humans share a common evolutionary history. When we talk about our ancestors we are referring to the ancestors of everyone on the planet.

Genes and clocks

Evolution occurs because of mutations (changes) that take place in the genes. In fact, these mutations are quite rare and the DNA of a species changes very slowly through the generations. Recently scientists have realized that they can use the number of differences between the DNA of two species as a clock, to measure the time that the species have been evolving separately from their common ancestor.

When the genetic clock is applied to human beings and chimpanzees, our closest living relations, the 2 per cent difference between our genes shows that we have been evolving separately for about 7 million years – about half a million generations: 7 million years ago an animal lived that was ancestor both to ourselves and the other modern apes.

To help you picture what this means, suppose your mother has a sister (your

▷ Many important discoveries of early human ancestors have been made in Africa by the Leakey family, Louis, Mary and their son Richard. These human footprints, preserved in fossilized mud, are 3.75 million years old. They were discovered by Mary Leakey in Tanzania in 1978.

aunt) and she has a daughter (your cousin). If you hold your mother's hand and your cousin holds her mother's hand, and then your mothers both hold hands with their mothers, they will both be holding hands with the same person – your grandmother. Your grandmother is a common ancestor of you and your cousin.

Now imagine standing next to a chimpanzee and doing the same thing. Two lines would form stretching back in time. You would stand at the head of one line, holding hands with your mother, who holds hands with her mother (your grandmother), who holds hands with your great-grandmother and so on. In the second line the chimps would do the same, each one holding hands with her mother back along the line.

Now imagine travelling along the lines to see what your ancestors looked like. As you pass by each person you step back in time by a generation. The period between human generations is about 20 years. Just five people down the line you are looking at someone from the last century; 50 people along the line you are passing your ancestor from a thousand years ago, 500 people down the line you see your ancestor from 10,000 years ago – before

RECORD IN THE ROCKS

The evidence for evolution lies hidden in the rocks. At the seaside, waves pounding against cliffs expose fossils in the rocks. These are the remains of once-living creatures buried for millions of years. Swanage in Dorset on the south coast of England is a popular destination for fossil hunters. Fossils are also found during excavations for road-building projects and at sites where erosion or land movements have exposed the rocks beneath the surface. For example, the Rift Valley in East Africa is an area rich in fossil sites.

The kind of rock in which fossils are discovered is called 'sedimentary' rock. Sediments are layers of mud washed into estuaries (river mouths) and lake beds by flowing rivers. As sediments build up, the lower layers harden under the weight of the layers above. Provided the layers have not been distorted or mixed up by geological processes, then fossils found in the lower layers must be older than those in layers above, since the lower layers were deposited first.

We can work out the age of sedimentary rocks by measuring the amount of certain radioactive materials they contain. These materials give out invisible rays that can be counted with a special detector. Older rocks are less radioactive than rocks formed more recently. This ability to date rock layers enables us to read the history of life from the fossils found in them.

Radioactive dating shows that meteorites (rocks that have fallen on to Earth from space) are 4600 million years old. This is also thought to be the age of our planet, which formed at the same time as the meteorites and the other planets in the Solar System. The Earth's surface is mostly much younger than this, however, because of the action of volcanoes, the weather and other rock-forming processes.

But the first evidence for life, remains left by bacteria and algae (simple plants), is in rocks 3400 million years old. About 1500 million years ago the first fungi appeared and the first evidence for protists (see page 9) is about 1000 million years old. The first animals, coelenterates (jellyfish, anemones and corals) and simple worms appeared about 600 million years ago. By 400 million years ago, land plants and fish had evolved. Reptiles moved from the sea to the land about 355 million years ago. Then, 250 million years ago, dinosaurs appeared at about the same time as the first mammals. But it was the dinosaurs who were more successful and for nearly 200 million years they were the dominant animal group. Then, 63 million years ago, the still unexplained extinction of the dinosaurs occurred and the age of the mammals began. Humans are mammals, and the fossil evidence suggests that our species is descended from early apes that appeared about 7 million years ago.

written history began. Your 500 times great-grandmother is still obviously a human being and, if you look across to the chimps' line, it is still clearly a line of chimps. As you continue walking faster and faster back down the lines past thousands and then tens of thousands of ancestors, you do not see any sudden changes, as each person in your line looks very like her mother and the same is true of the chimps.

Then gradually you realize that, although you were not aware of the change from generation to generation, your ancestors are looking different from you and your mother – they have more body hair, stand less upright and are starting to get smaller. Now they are resting their body weight forward on their hands. The further back you go the more similar the two lines become, until eventually, when you have passed about half a million generations, the two lines meet. Two small apelike animals, sisters, are holding hands with their mother. One sister starts the line that ends with the modern chimp. The other sister starts the line that ends with you. Their mother is a common ancestor of all chimps and all human beings on Earth.

Ancestors of modern apes and humans

10 million years ago

5 million years ago

Australopithecus (Lucy)

Homo habilis
The first tools

2 million years ago

Homo erectus in Africa

1 million years ago
First humans in Europe
(probably *Homo erectus*)

Homo sapiens in Africa
The first use of fire

500,000 years ago

Uncovering the past

200,000 years ago

Paleoanthropologists are scientists who try to trace the human family tree. They look along the line of descent from the apelike mammals that lived 7 or so million years ago to modern human beings. Their aim is to discover the stages through which our ancestors passed, and so understand how evolution by natural selection could make us into what we are today.

Homo sapiens sapiens (modern humans) originate in Africa

Homo sapiens neanderthalensis (Neanderthals) spread through Europe

100,000 years ago

Modern humans spread out of Africa into Asia and Europe, replacing other humans

In many ways these investigators are detectives as well as scientists. You cannot do experiments on the past. All you can do is search for evidence in the form of fossils and other remains, and hope that you can fit the evidence together like the pieces of a jigsaw to reveal a picture of human history. Unfortunately the jigsaw is very far from complete, and we do not have the picture on the box to guide us! Some pieces are missing and we are not yet sure where other pieces fit. The human family tree always causes much controversy and many heated debates, but the picture is starting to fall into place.

The first art

The first technological revolution

Homo sapiens sapiens spreads into North America

△ Through the painstaking work of paleoanthropologists, such as the Leakeys and others, we are now beginning to piece together the human family tree. The first human beings appeared in Africa 2 to 3 million years ago.

20,000 years ago

Humans reach Greenland and the most remote Pacific Islands

1000 years ago

PRESENT

From apes to humans

There have been few fossils found of our ancestors from 7 to 4 million years ago. An early ape called *Dryopithecus* may have been the ancestor of modern apes including modern humans. Some people have suggested that a descendant of *Dryopithecus* called *Ramapithecus*, which was apelike but had human-like teeth, was in the line that led to humans, though others think it was in the line that developed into orang-utans.

About 4 million years ago the first human-like creatures appeared – *Australopithecus*. Many fossils of these animals have been found in Africa, including an almost complete skeleton of a young female who has been christened 'Lucy'.

There were at least four species of *Australopithecus*. Lucy was just over 1 m tall, but other individuals were as tall as modern humans. *Australopithecus* was 'bipedal' – this means that Lucy walked upright on two feet. Richard Leakey believes that bipedalism was a key evolutionary change that freed apes' hands to use tools. In other ways, however, *Australopithecus* probably lived very like bands of apes today, perhaps spending some of their time in trees. Their teeth were suitable for eating roots and berries, and they probably searched for food together in large family groups. There is no evidence that they manufactured tools, but they might have picked up sticks and stones to use as tools or weapons.

Handy man

The first manufactured tools were simple lumps and flakes of rock. These were probably used for chopping plants and getting meat from carcasses. These tools were made by *Homo habilis*, the first-known humans. Remains of *Homo habilis* were found by the Leakey family at Olduvai Gorge in Tanzania in Africa. These remains are thought to be 2.5 million years old. *Homo habilis* skulls show that their brains were nearly twice the size of the brain of *Australopithecus*, but only about half the size of a modern human brain. *Homo habilis* had teeth suited to eating meat as well as plants, and they probably scavenged meat from dead animals.

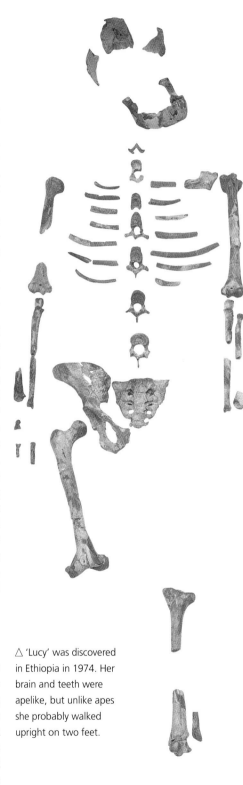

△ 'Lucy' was discovered in Ethiopia in 1974. Her brain and teeth were apelike, but unlike apes she probably walked upright on two feet.

Upright man

Homo erectus appeared 1.5 million years ago. They had brains nearly two-thirds the size of ours and similar, but sturdier bodies. *Homo erectus* were the first humans to spread out of Africa. Their remains have been discovered in Europe, the Middle East and Asia. There is good evidence that *Homo erectus* lived a wandering, 'hunter-gatherer' lifestyle. Groups of *Homo erectus* set up camps to use as bases for their food-gathering expeditions. They returned to the campsites with their kills to butcher them. Scattered among the bones they left behind are the stones and other materials they used to make their tools and weapons. Their tools were more carefully made than those of *Homo habilis*. Modern experiments have shown that much practice and skill is needed to make hand axes like those found at *Homo erectus* sites.

Wise man

The *Homo erectus* lifestyle survived for more than a million years. Then, about half a million years ago, more advanced humans – *Homo sapiens* – started to appear. Neanderthals (*Homo sapiens neanderthalensis*) were successful members of this species. They spread through Europe 130,000 years ago and survived until about 35,000 years ago. They are named Neanderthal after the Neander valley in Germany where their remains were first found.

Modern human origins

Did modern human beings (*Homo sapiens sapiens*) evolve from Neanderthals? One theory is that modern humans gradually evolved from different early forms of *Homo sapiens* all around the world, possibly including the Neanderthals. If this is true, we would expect today to see differences between people in different parts of the world, because of their different lines of descent. Superficially this seems to be the case – people in different countries have different 'racial' characteristics such as their facial features, skin colour and hair type. But genetic comparisons show that all humans are very closely related.

Racial differences are only superficial, formed by relatively recent adaptations to different environments, or simply by chance, for example the distinct eye shapes found in the East and West. If humans had really evolved in different places from different ancestors then their genetic sequence would be much more varied than it is.

NEANDERTHALS

Neanderthals fit the classic 'cave man' image of popular films and comics, but these presentations almost certainly do not do justice to their skills and intelligence. They had heavy bodies, low foreheads and large teeth, but their brains were just as big as ours today. They lived in caves and huts, and even constructed small stone shelters. They were clever hunters and toolmakers, making many different forms of stone and flint tools. We do not know if they used language or created art. But they were the first people to bury their dead, and this fact alone suggests that they had a culture in which they cared for each other.

▷ An artist's impression of a Neanderthal hunting party. Although a work of the imagination, it is based on real evidence of Neanderthal life found in Europe. The hunters have trapped a woolly rhinoceros in a camouflaged pit.

The Eve hypothesis

Genes change through the generations by mutation. The more generations there have been, the more differences there will be between the genes of individual people. When scientists examine human genes from around the world, they find that the genetic sequences of African peoples are most varied. This suggests that modern human beings, *Homo sapiens sapiens*, have existed longest in Africa. The further away human beings are from Africa, the less varied are their genes. It would appear that modern humans originated in Africa and gradually spread through the world replacing other *Homo* species such as the Neanderthals. When the genetic clock is applied to present-day humans, it seems that our origins can be traced back to a single female ancestor, 'Eve', who lived in Africa between 100,000 and 200,000 years ago. Eve's children gradually spread through Africa, then, perhaps 60,000 years ago, into Europe and Asia. For a short while modern humans and other human species may have been in competition, but by 35,000 years ago *Homo sapiens sapiens* had replaced all competitors, and was left as the only human species on the planet.

THE SPREAD OF HUMANITY

About 100,000 years ago human beings who looked like us appeared in Africa. Within a few thousand years, modern human beings replaced all other human species and spread throughout the world. Why are we so successful? The tools and the art of our ancestors help explain the key to our rapid progress – intelligence.

From 100,000 to 10,000 years ago the climate in northern Europe was arctic. This was the last Great Ice Age. The ice did not extend into southern England and France, but even there the climate was much colder than it is today. The landscape was one of cool grasslands and wooded river valleys. Wildlife was abundant. Herds of woolly mammoth (a kind of elephant), reindeer and bison grazed on the plains. There were rhinoceros and wild oxen, horses and goats, bears, wolves and sabre-toothed tigers. This was the environment into which *Homo sapiens sapiens* (modern humans) spread about 40,000 years ago, expanding northwards from the warm African homeland as their numbers increased. Food was plentiful in the new habitat, but it must have been a tough life coping with the climate, particularly the long cold winters.

Hunter-gatherers

For most of our history, human beings have been hunter-gatherers. The first modern humans, like previous species of *Homo*, lived by scavenging meat, hunting wild animals, and by gathering berries, roots and nuts. *Homo sapiens sapiens*, however, were more successful at this way of life than their predecessors. They used their adaptable minds to make better weapons and tools and more effective shelters.

At first, the new humans must have been in competition with other human species such as the Neanderthals – perhaps they fought, we do not know. There are some fossil skeletons that suggest there may have been some interbreeding between the two, though recent genetic evidence suggests this is unlikely. But, in time, the more efficient new humans came to dominate and by 35,000 years ago the Neanderthals, together with all other human species apart from *Homo sapiens sapiens*, were extinct.

Hunting and gathering is a very adaptable way of life. Nature provides all your needs, you follow wild herds, set up camp for a few weeks, then move on as the weather changes and the herds migrate to find fresh pastures. Humans have adapted this lifestyle to inhabit every kind of environment – from tropical forests, to grasslands in temperate (mild) zones, to arctic ice floes. The ability to adapt distinguishes *Homo sapiens sapiens* from nearly every other animal species, most of which are restricted physically to life in one particular habitat.

In some parts of the world the hunter-gatherer way of life has survived to the present day. The San people (bushmen) of the Kalahari desert in southern Africa live in small bands, each consisting of about ten families. The Kalahari is a dry environment, lacking in vegetation, in which knowledge and skills passed through the generations are needed to find enough to eat. When the food supply in one location has been exhausted, the band moves on. The San have lived in this area for thousands of years, but their territory has now

▷ These San people of the Kalahari desert require knowledge, skill and ingenuity to survive in such a harsh environment. Wild plants are their main food source. They also hunt small animals with poison-tipped arrows.

◁ There are few harsher environments than the Arctic wastes of northern Canada. Helped by some modern inventions, the Inuit people still follow the hunter-gatherer lifestyle that has enabled them to survive there for thousands of years.

been restricted by white settlers and many of the San have given up their old lifestyle to become farm labourers.

The Inuit people of Arctic North America are hunters too. Before adopting western technology, such as rifles and motorized vehicles, they fished, hunted seals and whales with harpoons (a kind of spear), and travelled by dog sled and kayak (a small canoe). The traditional Inuit way of life was highly adapted to survival in extreme cold. In summer Inuit families lived in tents covered with seal or walrus skins, while winter homes were built from stone and insulated with moss. On long winter hunting trips over the ice, the Inuit built igloos from ice blocks. Thick parkas, trousers and boots made from double layers of skins and furs helped to retain their body heat.

The lives of the first *Homo sapiens sapiens* in Africa were probably more like those of the San than the Inuits, though food would have been more plentiful. But as the new humans spread from Africa into Europe, through Asia and beyond, they were clever enough to devise new tools, make warm clothes and build efficient shelters, coping with challenges posed by hostile new environments.

Remains of ancient campsites give us insights into the hunter-gatherer lifestyle of the first 'modern' Europeans. At Pincevent on the banks of the River Seine in France, there is a site that was occupied by nomadic hunters at various times towards the end of the last Ice Age, 10,000 to 12,000 years ago. Excavations reveal that these people were reindeer hunters. All that remains of their camps are the hard materials such as stone and bone that do not decay.

The mammoth hunters at Mezhirich in the Ukraine had to cope with even colder conditions than the reindeer hunters of Pincevent. Excavations of a 14,000-year-old site on the River Dneiper have uncovered the remains of five houses built from mammoth bones. These houses may have been covered with mammoth skins. The people who lived here cooked on fires inside their houses and ate birds and fish as well as red meat. Bones of foxes and wolves, together with bone needles, suggest that some animals were hunted for their fur – probably to make warm clothing.

▽ Reindeer bones in stone fire hearths show that the hunters of Pincevent, in northern France, brought their kills back to camp, where the carcasses were butchered and cooked. Broken flints found around the fires show where the hunters sat to make their tools.

A TECHNOLOGICAL REVOLUTION

Homo sapiens sapiens was not the first human to make tools and weapons. *Homo habilis*, *Homo erectus* and the Neanderthals had all been toolmakers too, but their relatively simple tools changed little for hundreds of thousands of years. About 35,000 years ago, however, as modern humans spread, there was a technological revolution. Tools rapidly became more refined and specialized.

For most of human history, flint has been the most important toolmaking material. Flint is a very hard stone found as large lumps in chalk. When flint is struck, or knapped, with another stone, it chips or shatters, producing flakes with sharp edges. A lump of flint can be knapped fairly easily into a crude hand axe, but *Homo sapiens sapiens* developed flint toolmaking into a sophisticated technology. Ultimately, a skilled flint toolmaker could produce fine flint knife-blades, delicate barbed arrowheads and needle-sharp tools. In a Stone Age camp, many hours must have been spent around the fire, knapping stones into shape. Perhaps the younger members of the group watched, trying to copy skilled toolmakers as they worked.

The new humans soon learned how to use other natural materials like antler, bone and ivory. A finely made bone harpoon head, discovered in Africa, is 90,000 years old. It is perhaps the earliest indication that a new kind of *Homo sapiens* had emerged, capable of skilled craftsmanship. Then, as the new humans spread, carrying their skills with them, they began to decorate the things they manufactured with carvings, they made beads and necklaces to decorate their bodies, they carved stone and ivory figures, and they created the first works of art.

▽ The first human tools, made more than 2 million years ago, were little more than crudely chipped lumps of rock.

△ This beautifully made flint knife-blade shows the refined toolmaking skills developed by *Homo sapiens sapiens*.

The first artists

In December 1994 the archaeologist Jean-Marie Chauvet was exploring a cave in the Ardèche region, near Avignon in France. He climbed down a rope-ladder and shone his torch on to the cave wall. The light revealed images that had not been seen for thousands of years – horses, bison, bears and rhinoceros – painted with brilliant red, yellow and black pigments. Many prehistoric cave paintings have been discovered in France and Spain, but scientific tests show that those in the Chauvet Cave are much older than any found previously. These images could be 30,000 years old.

No evidence of painting, carvings or other images made by *Homo* species apart from modern man has ever been found. Art seems to have arisen with the spread of *Homo sapiens sapiens*. Some aboriginal art in Australia is now thought to be between 55,000 and 60,000 years old. Yet strangely, no art older than this has so far been discovered in Africa, where it is assumed that modern humans originated. This might be explained by the climate. Early humans in Africa probably spent more time in the open and had less need than the first Europeans to take shelter in caves. Their art was perhaps done outside, where it has long since been obliterated by the weather.

Magic and mystery

What does this early art mean? Like refined toolmaking, it signals a leap forwards in the intelligence of *Homo sapiens sapiens* compared to their predecessors. People who paint study the world around them and try to make sense of what they see. Art is often a way of expressing your ideas. Think about the drawings you have done. You probably had an idea or a story in your mind as you made your marks on the paper.

Some scholars have tried to read religious or magical meanings into cave paintings. In the Chauvet Cave, a bear's skull was found on a rock surrounded by other bear bones, as if it had been used in a ceremony. In other decorated caves, the footprints of young people have been found, which perhaps indicate that ritual ceremonies took place there.

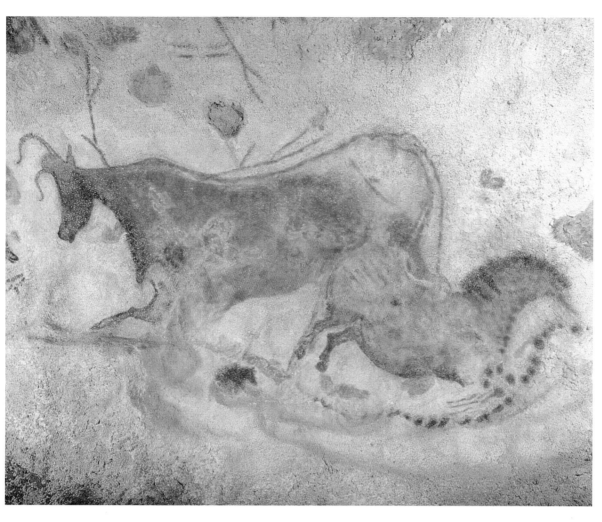

◁ Cave painting at Lascaux, France. *Homo sapiens sapiens* was the first, and so far the only species to use art to represent the world. Did the use of language begin at the same time? There is a lot of disagreement over this question.

But the truth is that we will probably never know exactly what the first art meant to our ancestors. Imagine archaeologists thousands of years in the future digging up a drawing of the Garden of Eden. If they had this as their only clue, it would be impossible to guess the story it represented. And yet it is hard to imagine that the vivid animal pictures on the wall in the Chauvet Cave are just decoration, a kind of prehistoric wallpaper. They have been painted by skilled artists in places that might have been important gathering points. Surely these images must have held deep meanings for the people who met under them?

The first talkers

Perhaps, above all else, it is our ability to talk – to invent and use language – which marks out *Homo sapiens sapiens* from all other animal species. When and why did human beings first talk? As with so many aspects related to early humans, we can only guess about their language skills – their voices died with them. Some people argue that the emergence of a complex language, which allows humans to tell stories about the past, to share their knowledge and experiences, and to plan for the future, is the explanation for the leap forward in Stone Age technology and the invention of art at some point in the past 100,000 years. It was as if with *Homo sapiens sapiens* human brain size had evolved to a stage at which language was suddenly possible.

Another theory is that language developed more gradually, right from the time that the first *Homo* species appeared several million years ago. This theory proposes that language became steadily more complex as human brains became bigger. Indeed, the evolution of larger brains may have been due to the increased chances of survival gained through language. The size of human brains increased steadily for more than 2 million years – some Neanderthals had larger brains than modern humans. The fact that brain size did not jump with the appearance of *Homo sapiens sapiens*, is taken as evidence for a gradual rather than a sudden evolution of language. Further evidence for this comes from the study of fossils of human skulls.

In most modern humans the ability to use language is located in the left side of the brain (see pages 86 to 91), which also controls the right side of the body. Nine out of ten people are right-handed. In contrast, modern apes use both hands equally and, although some attempts have been made to teach chimpanzees to communicate with language, their language facility is less than that of a three-year-old human child. It seems that left and right sides of chimpanzee brains are not specialized in the same way as modern human brains. Some of the earliest *Homo* skulls, however, do indicate a slight size difference between the left and right sides of the brain. Also, detailed studies of the earliest tools indicate that the majority were made by right-handed people. These facts suggest that language may have started to emerge more than 2 million years ago.

But if language had been evolving steadily for 2 million years, how then can we explain the appearance of art and the explosion of technology 40,000 years ago? On the other hand, if language did evolve very quickly, along with art, then what explains the steady increase in the size of *Homo* brains during the previous 2 million years? If Neanderthals were not talking, what were they doing with their large brains? The debate continues.

The first Australians

There has not been a direct land link between Asia and Australasia for at least 500 million years. During this time the plant and animal species of these regions have evolved quite differently. In Asia, for example, there are monkeys and elephants, whereas in Australasia there are koalas and kangaroos.

Modern humans started to move from South-East Asia into Australasia 50–60,000 years ago, following the string of islands known as the Indonesian Archipelago,

which stretches from the south-east tip of Asia towards the north-west of Australia. The journey must have been made by boat. The hunter-gatherer lifestyle of these settlers remained unchanged almost to the present day.

The first Americans

In contrast to Australasians, genetic studies suggest that most native North and South Americans are descended from just one small group that migrated across the Bering Strait from Siberia in north-east Asia 15–35,000 years ago. Perhaps this was a group of mammoth hunters pursuing their prey across the ice floes. Within a few thousand years, the descendants of these adventurous migrants (the Amerindians) had spread throughout the Americas, forcing many animal species, including the mammoth, into extinction – just as hunters did throughout Europe and Asia at about the same time. At later dates two other migrant groups followed the original Americans across the Bering Strait. About 6000 years ago the Athapascans made the crossing. Some of their descendants became the Apache tribe of the

American south-west. Then, about 4000 years ago, Arctic hunters, the Inuit and the Aleuts, started to spread across what is now northern Canada, finally reaching Greenland.

Island hopping

Perhaps the most adventurous migrants of all were the people who colonized the thousands of islands dotted throughout the vast Pacific Ocean. Imagine setting off with your family in a canoe or on a raft, to travel for weeks, with no idea where your journey would end. But where did the Pacific islanders come from? Did they travel east from Asia through Australasia, or had they come west from South America?

The Norwegian anthropologist and adventurer Thor Heyerdahl believed that the islanders' origins were in South America. In 1947 he set out from the coast of Peru on a wooden raft, which he called the Kon-Tiki, to prove that such a journey was possible. Landing in Polynesia after 101 days at sea, he proved that such a journey could have been made. But his theory was mistaken. Genetic tests now

---

THE SPREAD OF HUMANITY

From their probable origins in Africa, *Homo sapiens sapiens* (modern humans) spread throughout Europe, Asia and eventually the entire world. Natural barriers formed by mountains, oceans and the climate sometimes delayed progress, in places for tens of thousands of years. But humans lit fires and built shelters to cope with the cold. They also made warm clothing from animal skins, invented special shoes to travel across deep snow and eventually built boats to venture across oceans. Until recently the origins of different peoples around the world have been the subject of much debate, but genetic evidence now allows us to trace links that show how and when humanity spread.

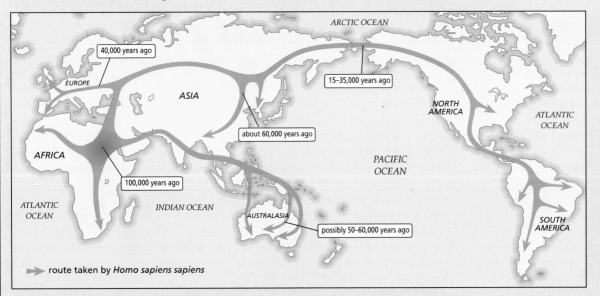

◁ 100,000 years ago modern human beings began to spread out from their African homeland. The arrows and dates indicate the directions and timings of their migrations, as they populated the world.

route taken by *Homo sapiens sapiens*

△ The remote islands of Polynesia are separated by vast stretches of the Pacific Ocean. The colonization of these islands by seafarers paddling tiny canoes was one of the great adventures in human history.

prove that the Pacific islanders are not related to the Amerindians of South America. The islanders' ancestors had travelled east, from Asia.

More than 20,000 years ago people known today as the Papuans spread from South-East Asia into New Guinea. A group known as the Melanesians moved on to occupy the islands north and west of Australia, including Fiji and Vanuatu. Then, about 3000 years ago, another group, known as the Polynesians, started to spread. During the following 2000 years the Polynesians colonized the most distant Pacific islands, reaching Hawaii and Easter Island in about AD 400, and finally colonizing New Zealand in about AD 900.

Recent arrivals

Like the Pacific islands, some other remote parts of the world were not reached by human beings until relatively recently. The European Norse or 'Vikings' were also great seafarers and reached Iceland in AD 870 and Greenland in AD 1000 (shortly after the Inuit, who had arrived in the opposite direction from North America). Viking sagas record that they also established settlements in 'Vinland'. We now know that Vinland is in North America

and the Vikings were the first European colonists of the 'New World' – 500 years before Columbus. The remains of a Viking village have been located on the coast of Newfoundland in Canada.

The first humans to lay claim to a particular place are rarely the last. As the human population has increased, people have continued to move from their homelands, often competing fiercely for territory and natural resources. Some of the greatest migrations took place following the 'discovery' of the New World in the 15th century by explorers such as Columbus and Cabot. In the 19th century in particular, millions of people migrated from Europe to set up new colonies in the Americas, Africa and Australasia, often with little regard for the rights and lives of aboriginal peoples of the lands they were moving to.

Today, although most of the inhabitable land on the planet is now occupied, people continue to move. Whether driven by famine, as people in Ethiopia were in the 1980s, conflict, as in Rwanda and the former Yugoslavia in the 1990s, or simply by the spirit of adventure and the desire for a better life, human beings are always ready to pack up their belongings and move on. Perhaps we are all still wandering hunter-gatherers at heart.

▽ The conflicts that have marked the 20th century have produced endless columns of refugees. Refugees are the innocent victims of wars and natural disasters. Fearing for their lives, these unwilling migrants abandon their homes and flee to an uncertain future.

BODY

The human body is an amazing living machine. Fuelled by the energy equivalent of a cupful of petrol a day, and supplied with a few basic raw materials, it grows, moves, repairs and reproduces itself, surviving for up to 80 years or more. How is it built and how does it work?

BUILDING A BODY

All human bodies are made to the same basic plan.
But within this plan there can be enough variety to make
each of us unique.

Imagine an alien biologist visiting Earth from another planet and attempting to write a description of the human species. The following is an extract from his logbook.

'...and now we come to the most successful animal species on planet Earth, the human being. I will start by describing the appearance of this complex life form.

'The structure of the body is remarkably similar to that of other mammals, for example the cat, the elephant and the chimpanzee. The body is symmetrical, with the left side an almost exact mirror image of the right. Four limbs are attached to the main body, or trunk, which contains most of the vital body organs including the heart, the lungs and the stomach. The head is supported by a short neck on top of the trunk and it has two eyes which face forwards. Air passes in and out of the body through two openings in the face called the nose and the mouth. By controlling the air leaving the mouth, the human being can produce an enormous range of sounds, from grunting and talking to screaming and singing. Sounds are detected with two ears on either side of the head.

'Human life is fuelled by food and water which are taken in through the mouth. Solid waste is excreted from an opening at the base of the trunk. Liquid waste passes through a smaller orifice just in front of this. In the female this opening faces downwards. In the male it extends into a forward-pointing tube, which is also a reproductive organ.

'Unlike dogs, horses and most other mammals, the human being walks upright on two legs. This frees the arms for other activities such as carrying babies, making objects, driving a car or using a computer. The hands at the ends of the arms are the most developed tools of any life form on Earth and they are specialized for grasping and manipulation. Each hand has four fingers and a thumb. When these are used together, they are capable of very delicate and precise actions, for example threading a needle, playing the violin or drawing with a pencil.

'Adult males are generally larger than females and have more body hair and less body fat. This gives the male body a more angular appearance. Males can be distinguished by visible external reproductive organs at the base of the trunk. Adult females have two mammary glands on the upper part of the trunk. Milk from these glands is used to feed babies.

'This basic body plan is common to all members of the species, but there is considerable variation in the size and appearance of individuals.'

The alien biologist's description should help visitors to Earth identify human beings from other animal species they might encounter. He has also noted how varied we are. On any city street you can meet adults with heights ranging from less than 1 m to more than 2 m, and weights from less than 35 kg to more than 150 kg. The tallest human ever recorded, Robert

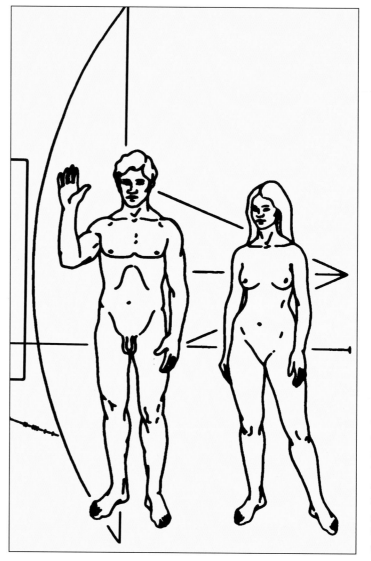

◁ This image of a man and a woman was carried aboard the unmanned Pioneer 10 and 11 spacecrafts launched in 1977. The picture also contains coded information that should help intelligent beings to locate our Solar System. It has been criticized for showing only apparently white people with the man taking the lead. Aliens finding it might also be surprised to discover that we normally wear clothes!

Wadlow from Illinois in the USA, was 2.72 m tall, and still growing, when he died in June 1940 aged 22. John Minnoc, a taxi-driver from Washington State, USA, weighed 442 kg in September 1976. That is as much as six average men. He died in 1983 aged 42. Abnormal heights and weights like these are caused by medical conditions which make the body grow too rapidly or put on excessive weight. Medical problems can also produce exceptionally small people. For example, a Dutch woman, Pauline Musters, was just 0.55 m tall and weighed 4 kg. She died in 1895 aged 19.

A rich variety

Some aspects of the variety among human beings can be explained by adaptations to the environment. The Inuit people of the Arctic, for example, tend to have relatively short stocky bodies. This shape helps them to keep warm by reducing the surface area through which heat can escape. Inuit hunters can grip harpoons at temperatures below those at which other people's

▷ The human body is more complex and adaptable than any machine.

fingers become numb. In contrast, people who live in hot dry climates, such as the Watusi of Rwanda and Burundi in central Africa, often have tall slender bodies to avoid overheating. Polynesians are among the largest, sturdiest people on Earth. This may be because only people with large reserves of stamina were capable of making the tremendous sea voyages to colonize the Pacific islands. Today Polynesians are often terrific sportsmen in games such as Rugby.

All around the world people have different skin colours. The reason for this variation has not been fully explained. Modern humans probably originated in a hot environment in Africa, and the first humans almost certainly had dark skins, like their African descendants today. A dark skin has the disadvantage of absorbing more heat than a light one, so there must be some other advantage that it provides. One possibility is that a dark skin prevents bright sunlight from destroying vitamins just below the skin's surface. As people moved into cooler climates, this protection was no longer required and lighter skins may then have evolved simply by chance.

Body building

What are little boys made from? Frogs and snails and puppy dogs' tails. What are little girls made from? Sugar and spice and all things nice.

These lines from an old nursery rhyme are intended to express more about traditional ideas of the personalities of boys and girls than about their biological composition. But sugar, spices, frogs, snails and dog tails would be perfectly good foods for both girls and boys. The human body is built from the same substances, or elements, as these foods. There are just over 100 different kinds of element. The most abundant elements in your body are hydrogen, oxygen, carbon and nitrogen. Many other elements are present as well, including calcium, phosphorous, sulphur, sodium and iron. Each element is made up of atoms.

height in metres

2.72
2
1.78
1.66
1
0.55

◁ Human beings come in a huge variety of shapes, colours and sizes. Despite this, every human body is built to the same basic plan.

Robert Wadlow   Mr Average   Mrs Average   Pauline Musters

Atoms are tiny particles of matter which link together to form all the living and non-living materials on Earth. A molecule is a unit composed of two or more atoms linked together. Two atoms of hydrogen and one of oxygen link up to make water, the most common molecule in your body.

More than half of your body is water. Water is a vital body component, used for dissolving and transporting other substances. You could survive for several weeks without food, but you would be dead within a few days without water to drink. It would be incorrect, however, to say that you are made from water. The substances that hold your body together, giving it shape and strength, for example in your muscles, skin and bones, are made from much bigger molecules based on the element carbon.

All living things are built from molecules based on carbon. Tens of thousands of different kinds of these molecules build our bodies, but they can be grouped into just a few basic types: lipids, which are fatty substances; carbohydrates, which include sugars such as glucose; the nucleic acids DNA and RNA, which carry the genetic instructions for the growth of the body; and proteins, which give the body strength and structure and act as chemical machines, for example to make muscles contract. Proteins are also cargo carriers, transporting smaller molecules such as oxygen through the blood and in and out of cells.

In theory, a human body could be built from a pile of carbon pencil leads, a few buckets of water, some nitrogen from the air and a few teaspoonfuls of other elements. But our bodies are not able to use these substances from scratch as food. We need something to combine them into molecules that we can use. Plants use the energy of sunlight to turn carbon dioxide and water into sugars. This process is called 'photosynthesis'. Plants can also extract elements from the soil through their roots to build the proteins and other molecules that make living things work. We survive by eating plants and other animals which have eaten plants and animals, to obtain the molecules that they have constructed. So we should eat sugar, spice, frogs and snails – they already contain the carbon-based molecules we need to build our bodies.

Living cells

The molecules of life, for example proteins and fats, are assembled in our bodies into a sequence of more and more complex structures. Cells are the basic production units, like tiny chemical factories. The activity of each cell is controlled by the DNA (see pages 12 to 13) held inside a small package called the 'nucleus'. Messenger molecules copy sections of the genetic instructions in the DNA code and carry it to parts of the cell called 'ribosomes'. These are construction units, which follow the instructions to manufacture protein molecules. 'Mitochondria' are the cell's power stations, releasing energy from nutrient chemicals supplied by food. In the 'endoplasmic reticulum' various chemicals the body needs are made, and waste products are broken down. The whole cell is surrounded by a thin wall called a 'membrane', which acts as a boundary and door keeper, selecting which substances can pass in and out of the cell.

Like different factories, cells do not all do exactly the same work. Although each cell contains all the genetic instructions for the whole body, only some of the instructions are active in each cell. The cells in your leg muscles contain instructions about the colour of your eyes, but only the muscle instructions are active. Different cells are specialized in this way to perform different jobs in the body. Red blood cells, for example, are packed with a special protein called 'haemoglobin' that can absorb oxygen. Muscle cells contain fibres that can slide over each other to make the cells change length. Nerve cells are extended into long thin fibres that can carry electrical signals.

Tissues, organs and systems

Imagine zooming into the body with an incredibly powerful microscope. At the highest magnification we would see how atoms are linked together into molecules. Then, if we started to zoom out again, decreasing the magnification in stages, we would see that the molecules are organized into the parts of cells.

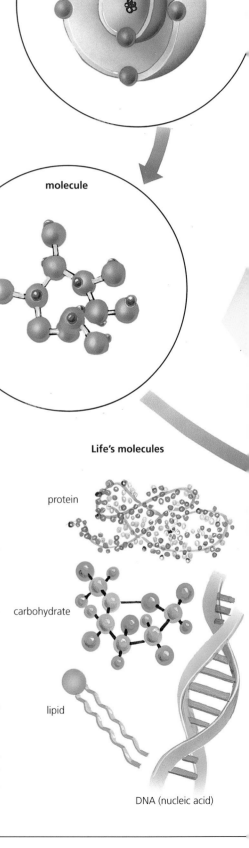

atom

molecule

Life's molecules

protein

carbohydrate

lipid

DNA (nucleic acid)

Building a body

organ (liver)

tissue

cell

mitochondrion

nucleus

endoplasmic reticulum

membrane

Zooming out further we see that similar cells group together to form tissues. For example, bundles of muscle cells make up muscle tissue. The cells are held together by connective tissue. This acts as both a framework and a route through which materials can pass to and from cells. Connective tissues consist of several components including 'collagen', a strong rope-like protein, and 'elastin', the stretchy protein that keeps your skin tight.

Zooming out one more stage, we see that tissues are grouped together to form organs such as the liver, the heart and the lungs. Organs are groups of tissues that work together to perform specific tasks. For example, the heart pumps blood.

Finally we see that the body's organs are organized into systems to operate the body's life processes. Our heart, blood and the blood vessels, for example, form our circulatory system (see pages 42 to 47).

◁ Understanding how a complex living thing like the human body is put together is rather like peeling the layers from an onion. Underneath each layer or level there is another layer to be discovered. Underneath the skin, our bodies are composed of organs such as the liver, brain, heart and kidneys. Each organ is built from tissues. Tissues are built from cells. There are different types of cells (see below). Cells are built from organelles. These are built from molecules, and molecules are built from atoms. But what are atoms built from? That is a question for a book on physics!

Types of cells

nerve cell

red blood cells

sperm cells

intestine cells

muscle cells

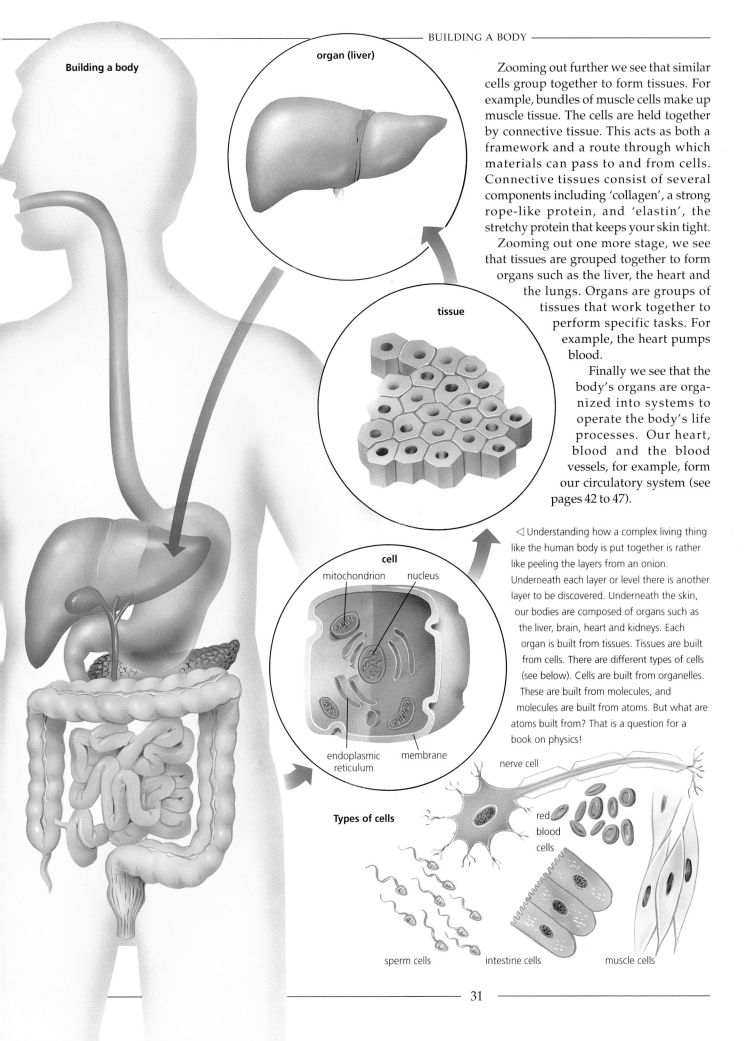

THE OUTER LAYER

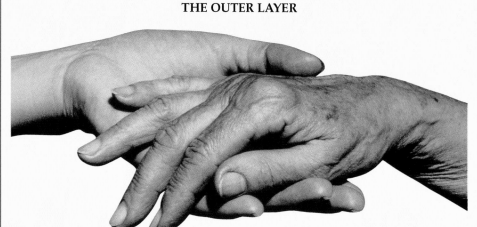

Just as individual body cells have an outer membrane to separate their inside from their outside, so your body as a whole has a waterproof, protective surface – your skin.

If you could peel off your skin and spread it out you would find you had a sheet of about 1.5 sq m. This varies in thickness from about 1 mm on the tops of the feet and backs of the hands, to several millimetres on the soles and palms. The skin weighs between 4 and 7 kg and is the heaviest organ of the body.

Skin is much more than just a waterproof cover. It is a defence barrier to keep out germs; it is a sense organ responding to temperature and various kinds of touch; and it helps to regulate temperature and protect the body from damaging ultraviolet rays from the Sun.

Under the microscope the skin is seen to be composed of different layers. The outer layer is called the 'epidermis'. It consists of flat dead cells packed with a tough protein called 'keratin'. This dead layer of skin is constantly being shed. The dust in our houses is proof of this – most of it is human skin. In fact, every year you shed about 4 kg of skin. In a lifetime you shed several times your own body weight!

Have you ever noticed how your skin becomes wrinkled when you lie in a soapy bath for a long time? This is because an oily substance called 'sebum', which makes the epidermis waterproof, has been washed away and your skin then absorbs water and becomes soggy. Sebum is produced by tiny glands at the base of hairs in your skin. The wrinkled effect is particularly noticeable on the palms of the hands and soles of the feet, where there are no hairs and the skin is thicker.

If you feel hot on a warm day or you exercise vigorously, you start to sweat. Sweat is a mixture of salt and water. It emerges from sweat glands through pores (tiny openings) in the skin. The evaporation of sweat from the skin cools the body, preventing it from overheating. When oil and sweat glands in the skin become blocked you may get spots.

Tanning is the skin's mechanism for protecting itself against the Sun's ultraviolet rays. A dark pigment (colouring substance) called 'melanin' is produced by special cells at the base of the epidermis. This pigment absorbs ultraviolet radiation from sunlight, to protect the tissues beneath. Exposure to sunlight increases the numbers of these cells, producing a tan.

Underneath the epidermis is the 'dermis'. This consists of the proteins collagen and elastin which make the skin strong but flexible – like a Lycra bodysuit. Running through the dermis there is a network of blood vessels. When you do hard physical work, you perspire and your face glows. Blood from inside the

body is passing through the blood vessels in the dermis so that heat can escape to the surroundings.

Your sense of touch is produced by various sensory cells, located in the dermis, which link to your nervous system (see pages 52 to 57). These detect pressure and temperature, and create the sensation of pain if the skin is damaged. Germs entering broken skin, for example through a scratch, are attacked by special cells from the body's defence system. Chemicals produced by these cells may make the scratch itch.

Fingerprints

The swirls and loops of fingerprints are unique to each one of us. In 1900 Edward Henry of Scotland Yard in London devised a system to identify criminals from fingerprints left at the scene of a crime. This system is now in use throughout the world.

▽ Skin is more than a waterproof wrapping for the body. Its blood vessels, glands and nerve endings sense our surroundings, control our temperature and defend us from germs.

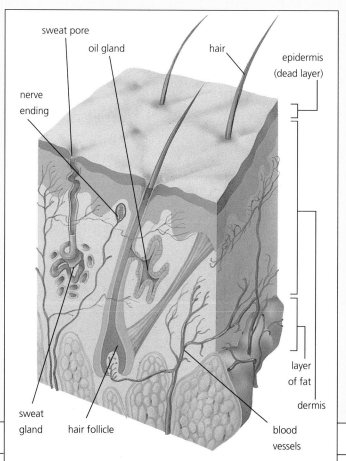

Hair and nails

Hair and nails are composed from a tough substance called 'keratin'. They grow from their roots throughout our lives, but only the root is alive and sensitive, so we can cut them without pain. A sharp tug on a nail or the hair is painful, however, as the force is transmitted to the roots. One of the functions of hair may be to extend our range of touch. Just as a cat's whiskers act as feelers in the dark, the hairs on our head give us an early warning that we are about to bang our head. Hair on our head also helps to reduce the loss of heat from our body.

Altogether you have about 5 million hairs on your body, the same number as a chimpanzee. But only the hairs on your head grow really thick and strong. They remind us that our ancestors had bodies covered with fur to help keep them warm. Now, only our goose pimples survive as a reminder of the time long ago when our ancestors fluffed up their fur against the cold.

Hair is coloured by melanin, the pigment which also produces your skin colour. Melanin-producing cells at the base of each hair colour it as it grows. Fair-haired people have fewer of these cells than dark-haired people. In later life the numbers of these cells may decrease. Your hair gradually turns grey and then white, as it grows without melanin.

Whether your hair is straight or curly depends on the shape of the follicle from which it grows. Round follicles produce round hairs which grow straight. Oval follicles produce flattened hairs which curl as they grow.

Human hairs grow at different rates, but on average the hairs on your head grow by 5 mm a week. If you never cut your hair it may eventually reach the ground, although it would probably stop at the waist. Members of the Sikh religion never cut their hair because of their beliefs.

The main function of the nails on the tips of our fingers and toes is to protect them. But they can also be used as tools, for example to play the guitar or to peel the skin from a grape. In other mammals claws are adapted for other purposes. Moles, for example, use their claws for digging, koala bears use them to climb trees and tigers use their claws as fierce weapons.

▷ If you never cut your hair it will probably grow to your waist. Diane Witt from America has the longest hair on record. In March 1993 her hair measured nearly 4 m!

▽ Fingernails grow by about 1 mm a week. Some people grow their nails long to look fashionable or to play an instrument that needs plucking, for example the guitar. They are not usually as long as this, however!

The human frame

A skeleton of 206 bones acts as a frame on which your soft body tissues like your muscles and skin hang. The skeleton is also a set of jointed levers which move your body around, and a cage which protects your delicate internal organs.

The bones of a newborn baby are quite flexible, so the baby can pull its feet into its mouth and suck its toes. The baby's bones consist mainly of 'cartilage', a flexible tissue built from the protein collagen. As the baby grows, the mineral calcium phosphate is laid down in the cartilage and the bones stiffen. In old age bones can become very fragile and fracture (break) easily. An elderly person may fracture their hip bone in a simple fall that would hardly affect a young person.

The longest bone in your body is the thigh bone, which in an adult can be 0.5 m long. The smallest bone is the stirrup bone inside the ear, which is about 3 mm long. Bone is remarkably strong. It can grow and repair itself when fractured. Bones are not solid. Inside they are filled with a spongy tissue called bone marrow. This has an important role in manufacturing blood and other body cells.

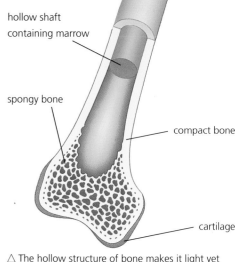

hollow shaft
containing marrow

spongy bone

compact bone

cartilage

△ The hollow structure of bone makes it light yet incredibly strong.

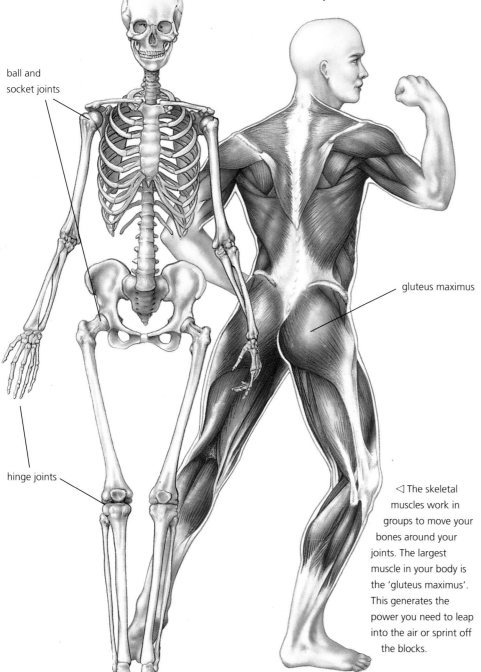

ball and
socket joints

hinge joints

gluteus maximus

◁ The skeletal muscles work in groups to move your bones around your joints. The largest muscle in your body is the 'gluteus maximus'. This generates the power you need to leap into the air or sprint off the blocks.

UNIVERSAL JOINTS

The individual bones of your skeleton are quite rigid, but where the bones meet they are linked by flexible joints, for example your knees and elbows. These allow different parts of your body to bend and change shape, rather like the arms and legs of a wooden puppet. Your bones are attached across the joints by flexible strings called 'ligaments'. The ends of bones that meet in joints are covered with smooth cartilage and separated by a layer of fluid. This allows them to move smoothly, reduces wear on the bones, and stops your joints from squeaking!

Some of your joints allow the bones to move more freely than others. Take the joints in your fingers, for example. You can only bend them forwards and backwards, whereas you can move your leg in many directions. Your finger joints are simple hinges that bend in one direction only, like the hinges on a door. But the hip joint that joins your leg to your body is a ball-and-socket that permits a much greater range of movements – from high kicks to the splits. In contrast, the movement between neighbouring vertebrae in your spine is relatively restricted, but by combining small movements of each of your 33 vertebrae you can bend your back quite considerably, like flexing the links of a metal watch-strap.

Muscle power

If the skeleton is the body's framework, then muscles are its motors. Just over a third of your body weight is muscle tissue. Our muscles produce the forces which both move our body around as a whole and operate its internal moving parts like the heart, lungs and digestive system. When you walk you are using 200 different muscles!

There are three distinct types of muscle. Skeletal muscle is attached to the skeleton and moves the bones around the joints. Cardiac muscle forms most of the heart and is different from all other muscles in the body. Smooth muscle (so-called because of its appearance under the microscope) is present in the walls of the various 'pipes' and 'tanks' inside your body like your stomach, bladder, veins and arteries.

Muscle fibres can only pull; they cannot push. When you bend and then straighten your arm you are using two separate muscles – one to bend the arm up and the other to straighten it. You can feel your biceps muscle on your upper arm contract (shorten) as you bend the arm. To straighten the arm again you contract the triceps muscle on the back of your upper arm and allow the biceps to relax.

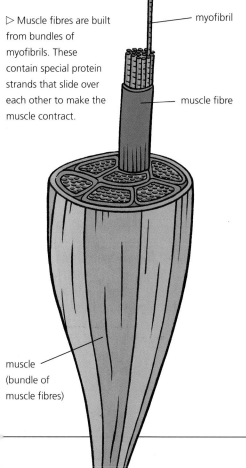

▷ Muscle fibres are built from bundles of myofibrils. These contain special protein strands that slide over each other to make the muscle contract.

myofibril

muscle fibre

muscle (bundle of muscle fibres)

▽ Many hours of training are needed to develop this kind of flexibility, balance and muscular strength.

▷ Lifting heavy weights increases muscle strength and size. Body builders develop massive biceps with exercises like this.

Skeletal muscles such as the biceps are attached to the bones by strong fibres called 'tendons'. You can feel the tendon in the crook of your arm as you flex your biceps. Your fingers are operated by tendons that run from muscles in your forearm through your wrist into your hand. You can see these tendons moving under the skin on the back of your hand when you make a fist.

Skeletal muscles are called 'voluntary' muscles because we can choose to make them contract. Our skeletal movements are under our conscious control. Cardiac and smooth muscles, however, are 'involuntary'. They work automatically by contracting regularly, to produce pumping actions to move or to restrict the motion of substances within the body. The heart muscles for example, work continuously for the whole of your life. They are the body's main pump (see page 45).

The digestive system also needs to move material around, and this is achieved by a pumping action called 'peristalsis'. Imagine moving a golf ball through a rubber tube by squeezing the tube just behind the ball with your hand. Food and waste are moved through the digestive system by a similar squeezing motion, produced by smooth muscle in the intestine walls.

The body in motion

Through muscle power alone the human body can move at 40 km an hour, lift 400 kg, or turn a triple somersault. These are the achievements of exceptional athletes following years of training. To reach these levels of fitness athletes must develop their muscles in appropriate ways. Our muscles contain at least two types of muscle fibres – slow fibres and fast fibres. The slow fibres can produce moderate forces for long periods without tiring. The fast fibres contract more quickly and with greater strength, but they soon grow tired. Athletes who require speed or strength develop fast muscle fibres by lifting heavy weights and training hard in short bursts. Endurance athletes, such as marathon runners, develop their slow fibres with steady training for long periods.

The basic plan for our body is programmed before we are born, but the way we live and exercise also influences our fitness, health and body shape.

NUTRITION

We are what we eat. Every mouthful is a rich mixture of the fuel and raw materials our bodies need to grow and maintain themselves. By choosing the right foods we can help our bodies perform at their peak.

Eating is a pleasure, but it is not just about satisfaction and enjoyment. At the end of a long cycle ride or a football game your body feels tired and hungry. You need to replace the energy you have used. But the food you eat is more than just fuel. Imagine a special fuel pump at the garage. When you fill your car from that pump, it provides not only the energy to run the engine, but also all the spare parts needed to repair any damage and the raw materials to make the car grow. With this kind of fuel your car would never need a mechanic and would eventually grow into a large truck!

◁ Almost everything we eat was itself once a living thing. The raw ingredients of a delicious meal contain all the nutrients we need to build and fuel our bodies. A mixture of plant and animal products, plus the occasional fungus such as a mushroom, provides us with the proteins, carbohydrates, fats, minerals and vitamins we require to maintain health and vitality.

Eat to live

The human body, like that of other animals, cannot build and maintain itself from non-living raw materials such as oxygen, carbon and hydrogen. Our foods are plant and animal products which already contain molecules similar to those in our bodies. After a meal, the process of digestion breaks down the food we have eaten and turns it into substances our bodies can use. If you were eaten by a lion, your body would be digested and rebuilt into parts of the lion's body in just the same way. Nature is a great recycler!

The substances we need from our food can be grouped into five main types – proteins, carbohydrates, fats, minerals and vitamins. Different foods contain different types and quantities of these nutrients.

Around the world people get their essential nutrients from a variety of different foods. In one culture carbohydrates may come from rice, and protein and fats from fish, beans or nuts. In another, bread and potatoes may provide the bulk of the carbohydrates, and meat and dairy products the protein and fat.

Proteins perform countless important functions in the body (see page 30). They are built in cells from molecules called amino acids. The human body requires just 20 amino acids to build all the thousands of different proteins it needs. These amino acids are obtained by digesting, or breaking down, the proteins contained in meat, dairy products, fish, beans, seeds, nuts and other foods.

Your body can convert some amino acids into others, so your diet does not need to contain all 20. But there are eight amino acids that we cannot make for ourselves. These are called the 'essential amino acids' and they must be present in our diet.

Carbohydrates are mainly used by the body as fuel. Starchy foods such as cereals, bread, pasta and rice, and sweet foods in the form of sugars, are a rich source of carbohydrates. Digestion breaks down starch and sugars into glucose, which is transported through the bloodstream to fuel the activity of our cells.

Your body needs fats to build the walls and internal parts of cells. Fat can also be used as fuel, and surplus fat is stored under our skin as a reserve supply of fuel.

△ Processing food changes its texture and makes it more digestible. This woman is pounding plantains.

contained must be replaced or you will become anaemic. You will look pale and feel weak. Dark green vegetables, treacle and liver are rich sources of iron.

Vitamins are molecules that your body needs in tiny amounts but cannot manufacture itself. Vitamin C is found in fresh fruit. Your body needs it to help make the protein collagen and to absorb iron. Without it the skin starts to flake, the blood becomes thin and the gums start to bleed. These are the symptoms of scurvy, the disease which was once common among sailors who spent long periods at sea with no fresh food.

Our bodies also need roughage, or fibre, which is provided by the tough, indigestible parts of plants. Vegetables, whole grains, beans and fruit are all good sources of roughage. Research has shown that, although fibre does not provide nutrition, it plays an important part in the action of the digestive system, bulking up your faeces (waste matter) and retaining water so that they are easier to pass. People who do not consume enough roughage are more likely to suffer from constipation, which can lead to more serious diseases of the intestines.

This is why we put on weight if there is too much fat in our diet. Butter, cream and cooking oils are familiar fatty foods. Fats are also 'hidden' ingredients in many processed foods such as cakes, chocolate, chips and hamburgers.

Minerals are elements such as calcium, phosphorous, iron, sodium and potassium that the body needs for building bones, carrying oxygen in the blood or controlling the amount and pressure of body fluids. You need to eat a lot of mineral-rich foods when your bodies are growing. An infant needs calcium from milk, for example, to build a strong skeleton. We all need a certain amount of salt in our diet to replace the sodium we lose by sweating, but too much salt can raise our blood pressure. If you lose blood then the iron it

Food as fuel

Have you ever set fire to a peanut? Peanuts are rich in oil and burn like tiny candles. The peanut candle is a good demonstration of food as fuel. When you eat a peanut, the chemicals produced by digestion are slowly 'burnt' in your body. In the process they release just about the same amount of energy as that released by the flame of the burning nut.

Scientists measure the energy content of foods by burning them inside a closed vessel called a calorimeter and measuring the heat produced. These experiments show that fats and oils such as butter and lard are the most energy-rich foods, containing 900 kilocalories of energy per 100 g of food. Carbohydrates and proteins contain about 400 kilocalories per 100 g. You can check the number of kilocalories in the food you buy in shops and supermarkets as it is indicated on the packaging.

Depending on what you eat, the energy needs of your body may be provided by carbohydrates, fats or proteins. Ideally most of the energy should be provided by carbohydrates, since it is wasteful for the body to break down protein and fat for fuel. Athletes and other sportspeople usually eat a diet rich in carbohydrates, particularly before a race or match.

◁ Only the seeds and the lettuce leaf are recognizable as parts of living things in this highly processed food product. You can soon put on weight eating easily chewed, energy-rich food like this.

COUNTING THE CALORIES

How much energy do we require in a day? This depends on our energy output. Lazing in the sun reading a book requires only a minute fraction of the energy needed to climb a mountain in a freezing blizzard. Arctic explorers may need two or three times as much energy to maintain their body weight as an average person. Most adults require somewhere between 2000 and 3000 kilocalories per day to maintain a steady body weight. A growing teenager needs a similar number of calories per day.

Energy input and output is a delicate balance. If your energy input from food is slightly greater than your output, you gain weight. If it is less, you lose weight.

▽ A resting human body needs about 1 kilocalorie of energy per minute (this equals 1440 kilocalories a day) to keep its basic life processes going. As our activity level increases, so do our energy needs. Vigorous sport consumes 10 to 20 times more energy than lying on the beach.

KILOCALORIES PER MINUTE

lying on beach: 1

working at a computer: 1.5 to 2

vigorous sport: 10 to 20

A good meal

Cooking and eating a meal can be one of life's great pleasures. Walking into a room when we are feeling hungry and seeing and smelling a beautifully prepared meal stimulates our senses and gets our digestive juices flowing. But why do we cook our food? No other animal does.

The most obvious effect of cooking is that it improves the taste of food and makes it more appetizing. Why this should be is probably explained by two other effects of cooking. Firstly, cooking begins the process of breaking down food, making it easier to digest. For example, humans find raw potatoes almost completely indigestible, but after they have been boiled for perhaps fifteen minutes they are easily digestible. The boiling breaks down the tough cell walls in the potato structure, releasing the starch from inside.

Secondly, cooking kills harmful microorganisms that may be living on food. This is particularly the case with meat. The warm pink interior of a poorly cooked hamburger is an ideal breeding ground for the bacteria which can give you an upset stomach.

But not all food should be cooked. Too much cooking can destroy the vitamins in fruit and vegetables. Fresh raw fruit and vegetables are healthy and very nutritious.

Once we have prepared and cooked our food, we are ready to sink our teeth into it.

Biting and chewing

The main functions of our teeth are biting, tearing, crushing and chewing food. Our teeth are adapted to our varied diet, so that the teeth used for cutting are at the front of the mouth and the crushing and chewing teeth at the back.

The part of the tooth you can see (the crown) is covered with a layer of enamel. This is the hardest material in the human body. The inside of a tooth is composed of dentine, a tough, shock-resistant material. At the centre of the tooth is a cavity filled with pulp, containing blood vessels and nerves. This extends from the root which holds the tooth in your jaw.

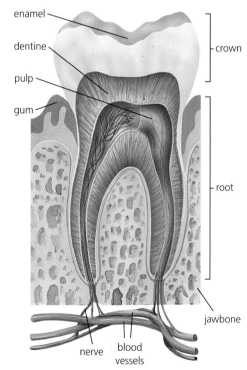

enamel
dentine
pulp
gum
crown
root
nerve
blood vessels
jawbone

△ A molar, one of the chewing teeth at the back of your mouth. Some nerves in your teeth are very sensitive to temperature. A hot drink followed by a cold ice cream can make your teeth ache.

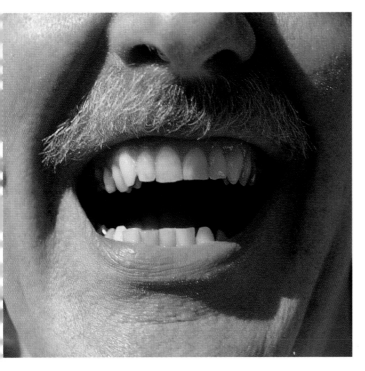

◁ We have two sets of teeth during our lives. The first set of 20 'milk' teeth starts growing when we are about 7 months old. When we are 6 years old adult teeth start growing and push the milk teeth out. By the age of 12 we have 28 adult teeth. The final 4 'wisdom' teeth at the back of the mouth do not usually emerge before the end of our teenage years. There are 32 teeth in a full adult set.

Digestion

Your digestive system is really just a long tube with your mouth, the entrance, at one end, and your anus, the exit at the other. If you laid out this tube in a straight line it would be about 7 m long.

During digestion food is broken down mechanically by chewing and by the churning action of your stomach. Then chemical demolition gangs called 'enzymes' get to work to break it down into the basic raw materials your body needs. Enzymes are protein molecules which speed up many different processes in the body. Once the demolition process is complete the products are absorbed into your blood for distribution to the cells, where they are stored and used. Any undigested wastes are expelled as faeces.

Digestion starts in the mouth when you chew food and mix it with saliva. Saliva contains an enzyme which breaks down starch molecules into sugars. This is why when you chew a piece of bread for several minutes it begins to taste sweet.

Tooth enamel cannot be replaced if it is damaged or worn. Bacteria feed on food particles trapped between our teeth and produce acid, which eats into the enamel to form cavities. These can give us toothache. Sugary foods and drinks are the worst enemies. Eventually the acid may eat away at the join between the crown of the tooth and the gum. The gum shrinks away, the tooth loosens and may fall out. If you visit the dentist regularly, pay attention to your diet and brush your teeth carefully, you can help to ensure that your teeth last you a lifetime.

Chewing your food mixes it with saliva to start the digestive process. The saliva moistens the food, and the combined actions of the teeth, lips, tongue and cheeks form it into a food ball or 'bolus'.

Swallowing

The first stage of the swallowing process, as you move the bolus to the back of the mouth and into the throat or 'pharynx', is voluntary. But once the bolus has entered the pharynx, automatic 'reflex' actions (see page 53) take over. These reflexes are stimulated by the moist saliva in the food, which is why it is so difficult to swallow dry unchewed food.

As you swallow, a flap called the 'epiglottis' seals off the windpipe to stop the food going down the wrong way. If it does, then you get a choking reaction to try to eject it. Normally the soft palette at the back of the roof of your mouth rises to seal off the exit to the nose. Sometimes, though, the gas from a fizzy drink escapes up your nose, producing a very strange sensation!

▽ This child will lose her first teeth as her adult teeth grow. But keeping her milk teeth healthy helps ensure that her adult teeth emerge straight and healthy too.

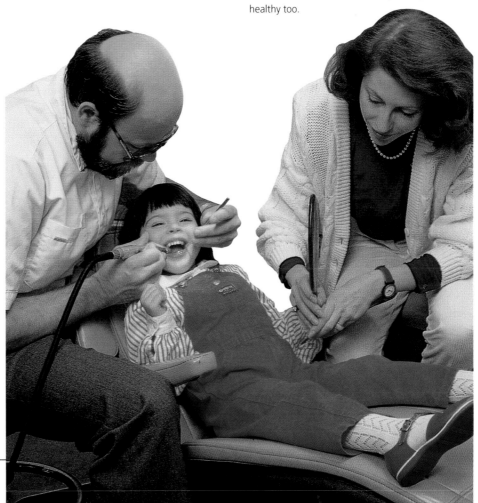

Once you have swallowed the food, it passes from your mouth into your stomach through a tube about 30 cm long called the 'oesophagus'. The stomach is the widest part of the digestive system and can expand to take all the food you eat during a meal. Inside the stomach, food is mixed with gastric juices. These contain hydrochloric acid which helps destroy bacteria, and enzymes which start the digestion of proteins. Sometimes some of the stomach acid finds its way back up into the oesophagus and we suffer from the burning sensation of indigestion or heartburn.

Food stays in the stomach for two to three hours and then passes, partly digested, into the small intestine, where most of the processing and absorption of the nutrients in food takes place. The small intestine (5.5 m) is by far the longest part of the digestive system. The partly digested food, now like a thick soup, is acted on by a cocktail of enzymes. Some of these enzymes are made in the intestine walls; others enter the intestine through a duct (channel) from the pancreas, together with bile. Bile is a dark browny-green liquid which is made in the liver and stored in the gall bladder. Its job is to break up fat globules which do not dissolve in water (rather like the action of washing-up liquid) so that they can be attacked by the fat-digesting enzymes.

The walls on the inside of the small intestine are covered with tiny finger-like extensions, called 'villi'. These contain blood vessels into which the digestion products eventually pass. The villi greatly increase the surface area of the intestine walls, making it easier for molecules to be absorbed into the blood. It has been calculated that the surface area of the villi inside the small intestine is equal to that of a tennis court. Some small molecules can pass through the villi walls unaided; others are helped by special carrier proteins.

After several hours in the small intestine undigested food, including roughage, passes into the large intestine. There are a lot of bacteria here which continue to digest food remains, often producing considerable quantities of gas. Moisture and minerals are absorbed from the mixture through the large intestine walls. As the contents of the large intestine move

▷ Food may take from a few hours to several days to pass through your 7 m digestive system. During this journey it is broken down into its basic building blocks by armies of digestive enzymes. Anything of value is absorbed into your bloodstream before the waste is excreted from your anus.

soft palette
pharynx
epiglottis
oesophagus
pancreas
liver
large intestine
stomach
small intestine
rectum
anus

down into the rectum, they are formed into faeces. Faeces consist of moisture, bacteria and undigested food in about equal amounts. They are coloured by bile. The faeces are stored in the rectum until they pass out of the body through the anus.

Storage

What happens to the nutrients which our body has extracted by digestion from what we eat? The blood carries them first to the liver. The liver is the body's main chemical processing plant with perhaps as many as 500 different jobs to do. One of these is to control the levels of nutrients in the blood, by processing, storing and releasing them to keep their flow even between meals. For example, excess glucose is stored in the liver as glycogen. When you have not eaten for a while the level of glucose in your blood starts to fall and you lack energy. This stimulates the liver to convert stored glycogen back to glucose and release it into the blood.

Food and health

All our food is derived from other living things. We have an amazing ability to stay healthy on incredibly varied diets, simply because all living things are composed of the same basic molecules. Once food has been digested, it makes no difference to your body if the glucose, fats and amino acids in its blood were once part of a cow, a cashew nut or a carrot. But eating the correct balance and quantity of food is important if we are to stay healthy. A balanced diet contains the right proportions of proteins, carbohydrates, fats, minerals and vitamins.

Food-related health problems in the West are usually caused by eating too much, by eating the wrong kinds of foods and by lack of exercise. Processed food provides large amounts of easily absorbed energy in the form of sugars and fats. It is all too easy to put on

KEEPING THE BALANCE

In wealthy societies most people should have no difficulty in finding sufficient quantity and variety of food to stay well nourished. A varied diet of bread, cereals, potatoes, fresh fruit, vegetables, nuts, fish, dairy products and meat contains all the nutrients the body needs. A vegetarian diet which includes eggs and some dairy products will also provide everything the body requires. Vegans, people who do not eat any animal products at all, should take care to supplement their diet with certain items, such as vitamin B12, that are not generally available from plants.

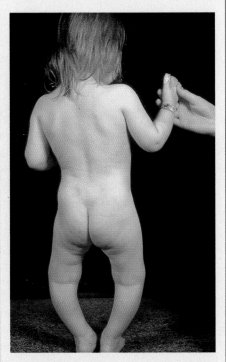

△ This child is suffering from rickets. Her bow legs are the result of soft bones, caused by lack of vitamin D.

to be poorly nourished. Anorexics do not eat enough to maintain a healthy body weight. They often have very negative feelings about their appearance and see themselves as much fatter than they really are. In extreme cases anorexics eventually starve themselves to death. Bulimics can appear to eat normally, but may binge secretly on comfort foods such as chocolate and cake. To keep themselves thin they make themselves vomit after eating. Repeated vomiting produces dehydration and loss of minerals which may upset the function of the heart. Anorexia and bulimia are psychological problems requiring medical help.

In stark contrast to overconsumption in the West, the problem confronting many people in developing countries is a lack of food – starvation. In people suffering from starvation, the body first makes use of all its fat reserves, then it starts to consume its own proteins to provide energy and the muscles waste away. As much as 60 per cent of the world's population have to struggle to survive on diets short of energy, protein, minerals and vitamins. These deficiencies create many health problems, slowing down the growth of children and reducing life expectancy by many years.

The world's agriculture is capable of feeding the present world population, but whilst there are food surpluses in the West there are shortages in many developing countries. Food and nutrition is a political and social issue as well as a scientific one.

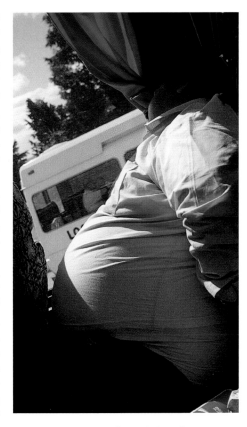

△ Obesity is the result of an imbalance between food intake and energy output. In the western world overeating of processed foods, combined with lack of exercise, means that many people are now obese. Health problems associated with obesity include heart disease and diabetes.

▽ Sailors in the British Navy were once given lime juice, a rich source of vitamin C, to prevent them from developing scurvy, as their diet lacked fresh fruit and vegetables. They became known as 'Limeys'.

excessive weight by eating too many sweets, hamburgers, chips and milkshakes. These foods do not contain as much roughage as unprocessed products and this may create bowel problems. Excessive fat in our diet, especially when combined with smoking, is associated with heart disease.

The eating disorders anorexia and bulimia which affect some young people, girls more often than boys, cause the body

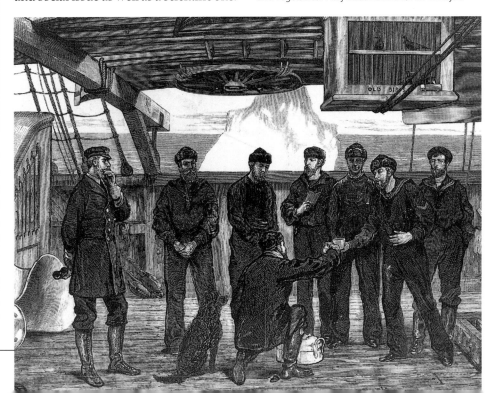

The Circulation

Your blood is a rich, life-giving cocktail. Every second of your life, the steady beat of your heart sends it on a never-ending journey around your body's circulation system.

The circulation of blood around your body acts as a transportation and communications system. Just as goods and mail are distributed by a network of roads and railways, so the blood carries chemical supplies, messages and waste products through the major and minor routes of the blood system. Body cells need regular deliveries of energy and raw materials. Waste products must be transported away and disposed of before they build up to poisonous levels. When tissue is torn or infected the problem must be detected, identified and dealt with by bringing in special cells to attack invaders and to repair damage. Excess heat from internal organs must be absorbed and transported to the skin, where it can escape.

Blood does all these jobs and more. A beating heart is such a clear sign of life that in the past, when their true functions were not understood, the heart and blood were thought to be responsible for feelings and emotions as well as physical health. A fiery and aggressive person was seen as 'hot-blooded', a cool and distant one as 'cold-blooded'. A kind and generous person was called 'warm-hearted', an unsympathetic and cruel person was 'cold-hearted'. We know now that these personality characteristics are a function of the brain, not the circulation, but we still use these evocative descriptions.

Blood's story

In its journey around your body the blood makes collections and deliveries in particular organs and tissues.

In the lungs blood collects oxygen and delivers carbon dioxide, a waste product of respiration (see page 49), which we breathe out. Blood delivers the oxygen and glucose needed for energy to the muscles and other tissues, and collects their waste products. In the walls of the intestine blood absorbs the products of digestion. It delivers these to the liver for storing and processing, and delivers the waste products to the kidneys for disposal in the urine. In glands such as the pituitary and thyroid glands the blood collects special molecules called hormones (see pages 58 to 61). These control various body functions such as growth and the rate at which your cells burn fuel.

▷ Arteries (red) carry blood from the left side of the heart to supply your body with oxygen. Veins (blue) return the blood to the heart. The arteries and veins are linked by a vast network of fine capillaries, which circulate the blood through all your body tissues.

brain

heart

lungs

vein

artery

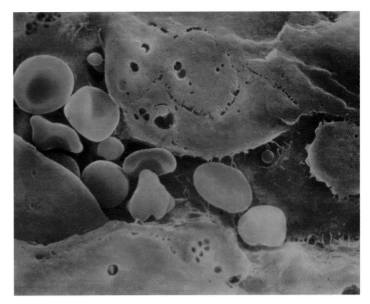

◁ When blood is spun at high speed in a machine called a centrifuge it settles into layers. The clear liquid at the top of the tube is called plasma. The dark layer at the bottom consists of tightly packed red blood cells. White blood cells and platelets collect in a thin layer between the plasma and the red cells.

Blood is thicker than water

Blood is not a simple liquid like oil or water. It is made up of tiny red and white cells, and small particles called 'platelets', floating in a watery fluid called 'plasma'. Each of these has a vital role to play in maintaining our bodies.

The red cells transport oxygen around the body. They give blood its colour. If blood contains fewer red cells than normal the patient is described as anaemic. Anaemia can be caused by lack of iron in the diet, by blood loss, and by various diseases such as leukaemia which affect the production and survival of our blood cells.

White blood cells play an important role in defending the body against infection. There are various kinds of white cells with different jobs to do. 'Lymphocytes' detect bacteria and other harmful invaders which carry foreign molecules called 'antigens'. The lymphocytes produce special proteins called antibodies that stick to the antigens to label the invaders. The labelled invaders are then attacked and engulfed by other white cells called 'phagocytes'. The pus which weeps from an infected wound consists mainly of white blood cells that have died defending the body.

Platelets are the circulation's self-repair kit. When you cut yourself and bleed they help to make the blood clot (set) and stop the wound from bleeding. Plasma is mostly water, in which various food substances, hormones and other important chemicals are dissolved and transported round the body.

△ Red blood cells are made almost entirely from a special protein called haemoglobin. Each haemoglobin molecule contains four iron atoms which capture oxygen molecules inside the lungs and then release them wherever they are needed in the body.

▽ When you cut yourself blood leaks from broken blood vessels. Platelets plug the gap and thin threads of a substance called fibrin make a web. These trap water and cells to form a clot. Meanwhile white blood cells attack invading germs. The clot hardens into a scab and new skin grows underneath.

A blood test is one of the most powerful ways that doctors can use for detecting infection and checking your general state of health. The doctor draws a syringe full of blood from a vein in your arm and the concentrations of the red cells, white cells or other components are checked. For example, if a blood test shows increased numbers of white cells in the blood this indicates that your body is fighting an infection. Checks might also be made to see if there is too much sodium from salt, or if the level of cholesterol (a kind of fat molecule) is too high. High blood cholesterol is thought to increase the risk of heart disease.

fibrin fibres white blood cell
red blood cell
'invaders'

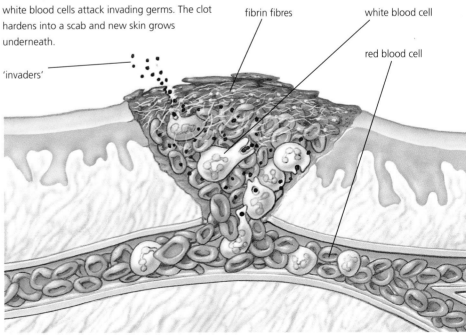

Blood supply

Our blood cells are replaced throughout our lives. More than 2 million red blood cells are destroyed and replaced every second. New blood cells are manufactured in our bones, in the soft spongy tissue called marrow. Cells in the marrow, called stem cells, divide and develop into the different cell types in the blood. The new blood cells enter the circulation by squeezing through the tiny thin-walled blood vessels which run through the marrow.

The lifespan of blood cells varies enormously. Some phagocytes survive for just a day or two. Lymphocytes can live for many years. The average lifespan of red blood cells is about 120 days. Old or damaged red blood cells are broken down in the liver, spleen and bone marrow. Most of their iron is recycled as new blood cells are manufactured.

Marrow which is making blood cells is red. Babies have red marrow in all their bones. As we grow up the marrow in many of our bones ceases its activity, fat is deposited and the marrow turns yellow. Only the marrow in the flat bones of the pelvis, chest and shoulder blades continues producing blood cells once we are adults.

Blood groups

Your body can quickly replace blood lost from a minor cut. But it cannot cope with a sudden large loss of blood, in a serious accident or during a major operation. In these cases a transfusion, a process by which blood from somebody else (a donor) is given to the patient, may be the only way to make up the loss fast enough to keep the patient alive.

Great care must be taken when introducing blood from a donor into a patient. The blood must be carefully checked to make certain that it does not contain infections such as the HIV virus that causes AIDS. It must also be tested to see if it is compatible with the patient's blood group. Your blood belongs to one of four main types or blood groups - A, B, AB and O - depending on the presence of different molecules on the surface of your blood cells. Only blood from a donor with the same blood group as the receiver can be

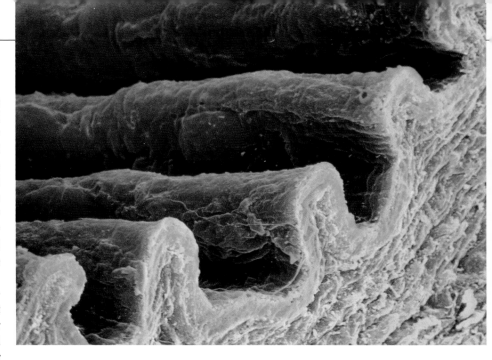

safely transfused in large amounts. Small transfusions of certain other groups may also be made without danger. If the wrong blood type is transfused into a patient, their white cells may react to the strange cells as an infection and attack them. This can be fatal.

As well as belonging to one of the ABO blood groups your blood is either Rhesus positive or Rhesus negative. This depends on whether you have a certain molecule on your blood cells. If a woman with Rhesus negative blood has a baby whose father is a man with Rhesus positive blood, then the baby may have Rhesus positive blood too. This can be very dangerous because as the baby develops in her womb the mother's white blood cells react to the baby's blood as if it was an infection. This reaction is not usually serious the first time the woman

△ This photograph shows a small section of an artery wall. Arteries have thick springy walls to withstand the pressure of the blood pumped from the powerful left side of the heart. The artery expands and contracts as blood pulses through it.

has a Rhesus positive baby. If she has a second Rhesus positive child however, her blood is already full of Rhesus positive antibodies from the first pregnancy and the reaction is very dangerous for the baby. The risk for the baby can be reduced with injections to stop the mother producing the antibody.

Blood vessels

An intricate network of tubes maintains a constant flow of blood around your body. Laid end to end they would go around the world at least twice. Arteries are the main highways supplying blood rich in oxygen from the left side of your heart to your body. Your arteries come close to your skin surface at your wrist and in your neck, so this it where it is easiest to feel the flow of the blood, your pulse. The arteries gradually divide into smaller and smaller branches. The smallest arteries are called 'arterioles'. The arterioles branch in turn into a network of tiny blood vessels called 'capillaries'.

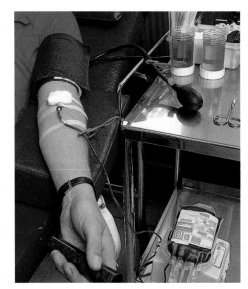

◁ Blood flows through a needle inserted into a vein in a donor's arm. Details of the donor's blood group are carefully recorded on the bag. The donor's body takes two to four weeks to make new cells to replace the donated blood.

Oxygen and nutrients are delivered to your body cells through the capillary walls. Capillary walls are very thin - just one cell thick - and even fairly large molecules can pass through them quite easily. Waste products pass through the walls in the opposite direction to be transported away by the blood. The capillaries gradually join up with each other to form large vessels called veins. The blood flows back through the veins to the right side of the heart, which pumps it into the lungs. The blood collects oxygen in the lungs and then returns to the left side of the heart, completing its circuit.

Heartbeat

A visit to the doctor usually involves some basic checks which quickly reveal if your heart and circulation are in good working order. The most straightforward test, and one that you can try for yourself, is to measure how fast your heart is beating - your pulse rate. Feel for the pulse in the arteries in your wrist or neck and count the number of beats in a minute. A typical resting heart rate is about 70 beats per minute.

Your heart rate rises during exercise because your active muscles require extra oxygen and food deliveries. During a race an athlete's pulse may reach 180 beats per minute or more, as the heart pumps blood around their body five times faster than when they are resting. With training, the heart can increase the amount of blood pumped at each beat. Very fit athletes can have resting heart rates of fewer than 50 beats per minute. Their hearts pump more blood with each stroke than those of untrained people.

Your heart rate is controlled by a combination of signals from the nervous system and hormones in the blood. When you work hard physically the brain sends nerve signals to the heart to beat faster. If you have a sudden shock, or are excited, angry or afraid, your heart rate may also increase. These emotions make the brain send signals to the adrenal glands which respond by releasing the hormone adrenalin into the blood. Adrenalin has the effect of raising your heart rate and making you breathe faster, so that your body is prepared for a fight or preferably a quick escape!

HEART OF THE MATTER

A healthy human heart beats approximately once a second. That is 3600 times an hour, 100,000 times a day, 30 million times a year, or more than 2 billion times in an average lifetime. Each time it beats, the heart pumps approximately 100 cubic cm of blood - about half a cupful. In a lifetime 200 million litres of blood flow through a heart - enough to fill more than twenty swimming pools!

An adult heart is about the size of a grapefruit. Located in the centre of the chest, it is really two separate pumps side by side. The right one pumps blood to your lungs where it collects oxygen. The left one takes this blood from the lungs and pumps it around the body to supply oxygen to the tissues.

The heart walls are made from special cardiac muscle which, unlike skeletal muscle (see page 35), does not become tired through constant use - after all your heart cannot stop for a rest!

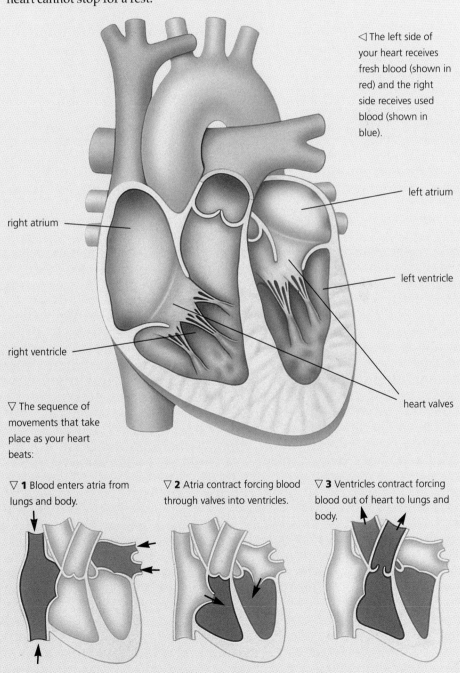

◁ The left side of your heart receives fresh blood (shown in red) and the right side receives used blood (shown in blue).

left atrium

right atrium

left ventricle

right ventricle

heart valves

▽ The sequence of movements that take place as your heart beats:

▽ **1** Blood enters atria from lungs and body.

▽ **2** Atria contract forcing blood through valves into ventricles.

▽ **3** Ventricles contract forcing blood out of heart to lungs and body.

A heart rate that is high may have several causes: the patient may be stressed or they may have a fever, which causes the heart to circulate blood more rapidly to carry away excess heat. If the patient is old their heart rate may be higher because their blood vessels have become narrower, making the heart work harder to pump blood around the body. If the pulse is very low or irregular then there may be a problem with the nerve signals that control its beats.

Heart sounds and signals

As well as taking your pulse the doctor may listen to the sound of your heart through a stethoscope placed on your chest. They would be able to hear your heart making two distinct sounds at each beat. The first sound, lub, is made by the rush of blood when the left ventricle contracts. This is followed by a quieter sound, dup, as the ventricle becomes wider again, and the valve snaps closed to stop the blood from flowing backwards. Leaky heart valves make a murmuring sound as blood is squeezed backwards through them. Many heart murmurs are

not serious at all, though some require the valves to be replaced with transplanted or artificial valves.

Electrical pulses from muscle cells in the heart trigger its regular pumping. These electric signals can be monitored with electrodes, metal discs attached to wires, stuck to the skin. The signals are plotted to produce a chart called an ECG (Electrocardiogram). The shape of the ECG can indicate if a patient has heart problems.

Feel the pressure

Pressure is needed to pump a liquid such as blood along a tube. The longer and narrower the tube, the higher the pressure required. Blood pressure is not the same in all blood vessels. The highest pressures are developed in the arteries that carry blood from the left side of the heart around the body. This pressure is needed to force blood through the network of narrow arterioles and capillaries. This explains why the left side of the heart is bigger and stronger than the right, and why your heart beat sounds and feels stronger on the left of the chest.

Arteries need thicker, more muscular walls than veins to withstand the higher pressure. If an artery is cut, bright red blood spurts out under pressure, so the blood loss is rapid and can be life-threatening. Cutting a vein is less dangerous. Lower-pressure, dark red blood seeps out more slowly.

Emergency!

A middle-aged man suddenly puts his hand to his chest and cries out in pain. He is very pale, cannot catch his breath and tells his wife that it feels as if his chest and arms are being gripped tightly. Fortunately she recognizes the symptoms of a heart attack and immediately telephones for an ambulance. Within a few minutes the ambulance has arrived, but by now the man has collapsed on to the floor. His heart has stopped beating regularly and is just fluttering weakly - if the emergency team do not act immediately he will soon die.

Working quickly the medics undo the patient's shirt and place two electrodes connected to a special generator against his chest. A button is pushed on the

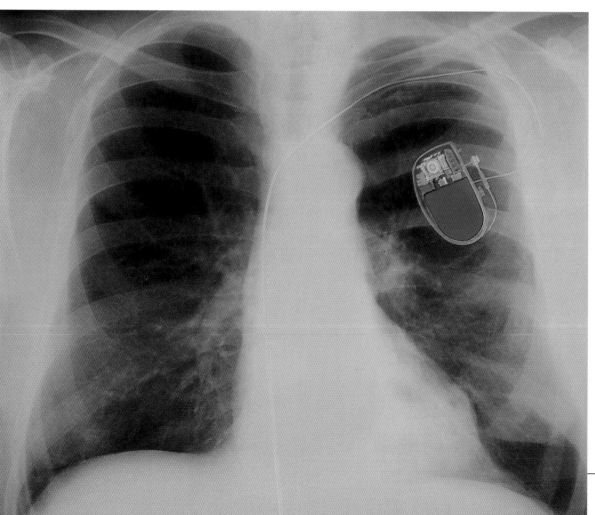

◁ Normally, special muscle cells in the heart produce regular electrical pulses that stimulate it to beat, but in some people with poor heart conditions these pulses get blocked. This patient is fitted with a pacemaker. This battery-powered device produces tiny electrical shocks to keep the patient's heart beating regularly.

PUMP IT UP

Your blood pressure is another check that a doctor can make on your heart. It is measured by pumping a sleeve full of air on the upper arm until it cuts off the blood flow to the lower arm. The doctor then listens for the sound of blood flow in the artery in your lower arm and notes the pressures at which sounds first appear and then disappear again, as the air is released from the sleeve. The blood flow only makes a sound when it is partially restricted by the sleeve - a bit like the noisy air flow from the narrow neck of a toy balloon. No sounds are heard when the blood is flowing freely or when it has ceased flowing completely. Many factors can alter our blood pressure. Loss of blood through an injury lowers it. Too much salt in the diet can raise it, as can stress and hardening of the arteries with age.

▽ Measuring your blood pressure is a routine part of a health check. A higher blood pressure than normal is called 'hypertension'.

and the heart muscles start to die.

Fortunately, in recent years the number of heart attacks in many countries has begun to fall. This is because research has warned people of the various risk factors that make heart disease more likely. The research shows that eating too many animal fats (found in red meat and dairy products), smoking and high blood pressure all increase your chances of developing heart disease. By watching what we eat, taking regular exercise and avoiding smoking, we reduce greatly our chances of suffering a heart attack later in life.

machine. The man's body jolts as a pulse of electricity passes through it, but then relaxes as his heart starts to beat normally again. The medics inject drugs to help increase the flow of blood to the heart muscles, and take their patient to hospital.

More people die of heart attacks in the western world than from any other single cause. A heart attack such as the emergency described above occurs when the blood supply to the heart muscles is interrupted. Without their supplies of oxygen and fuel, the heart muscles cannot function. As soon as the heart ceases to beat, oxygen supplies to the rest of the body are cut off too, and death follows within a few minutes.

The blood that supplies the heart muscles flows through the coronary artery, located on the outside of the heart. In people who suffer from heart disease, this artery becomes clogged with fatty deposits. These deposits restrict the blood flow to the heart. The victim suffers pain in the chest and arms and shortness of breath, if they do anything that raises their heart rate. A heart attack may occur when a blood clot, called a thrombosis, forms on the deposits, or a piece of the deposit breaks away and blocks the artery. In both cases the blood flow to the heart is cut off

▷ A healthy diet helps prevent heart disease. High fat foods such as sausages, chips, fried bacon, chocolates, cream and cakes should only be eaten occasionally. Eating fresh fruit, brown bread, vegetables, fish and lean meat will help keep your arteries clear and your figure trim!

BREATHING

A newborn baby takes her first breath, filling her lungs with air. A moment later she cries, making her first sound. Every few seconds for the rest of her life she will breathe in fresh air and then breathe out to get rid of waste gases and to laugh, scream, talk and sing.

Can you swim under water? It is a wonderful experience to swim under water like a fish. But, unlike a fish, you must soon return to the surface to fill your lungs with air. Without air you will die in a few minutes. Air contains the vital element oxygen, which your body needs to combine with sugars in your cells to release energy. The process of breathing in air, extracting oxygen from it, and using it to 'burn' food in your body, is called respiration.

Nearly all living things combine oxygen with food in this way to supply the energy needed for their life processes. A fish can survive under water because of its gills. Air dissolves in water: when you boil water in a saucepan you can see bubbles of dissolved air being released from the water as the temperature rises. The fish's gills are adapted to allow the dissolved oxygen in the water to pass into the blood which runs through the gills. Unfortunately our lungs are not capable of absorbing oxygen from water in this way. If we fill our lungs with water we drown.

The lungs of land-living animals, like dogs, monkeys, birds and ourselves, extract oxygen from the air of the Earth's atmosphere. When we take a breath and fill our lungs with air, oxygen passes into our blood. There it is collected by the red blood cells and distributed throughout our body by the circulation (see pages 42 to 47).

Breathing in

Breathing is an automatic action. We do not have to think about it. We each have two lungs in our chest protected by our rib cage. The right lung is larger than the left lung. When you breathe in (inhale), the diaphragm muscle, which lies between your stomach and your lungs, moves down and your ribs rise to increase the size of the chest cavity. This has a similar effect to pulling back the plunger on a bicycle pump to draw air inside. As the chest expands, air is drawn into the lungs through your nose and, if it is open, through your mouth.

The inside walls of your nose and the airways through which the air travels to your lungs are lined with hairs and special cells that make a sticky liquid called 'mucus'. The hairs and mucus together trap particles of dust and dirt in the air that you breathe in. The mucus also moistens hot dry air to prevent the lungs from drying out. When you have a cold too much mucus is produced and your nose becomes runny or blocked. This makes breathing more difficult.

From the nose, the air passes into the throat (pharynx) and then into the windpipe (trachea). About 20 cm down, the windpipe divides into two branches called 'bronchi'. If you suffer from bronchitis then you have an infection of the lining of the bronchi, which causes them to produce too much mucus and gives you a chesty cough.

◁ The human lungs can expand to hold 4 to 6 litres of air, nearly a bucketful. In normal breathing we use only about a tenth of this capacity, but trained opera singers like Jessye Norman use their lungs to the full. A deep breath provides the air to sing a musical phrase which can fill a concert hall.

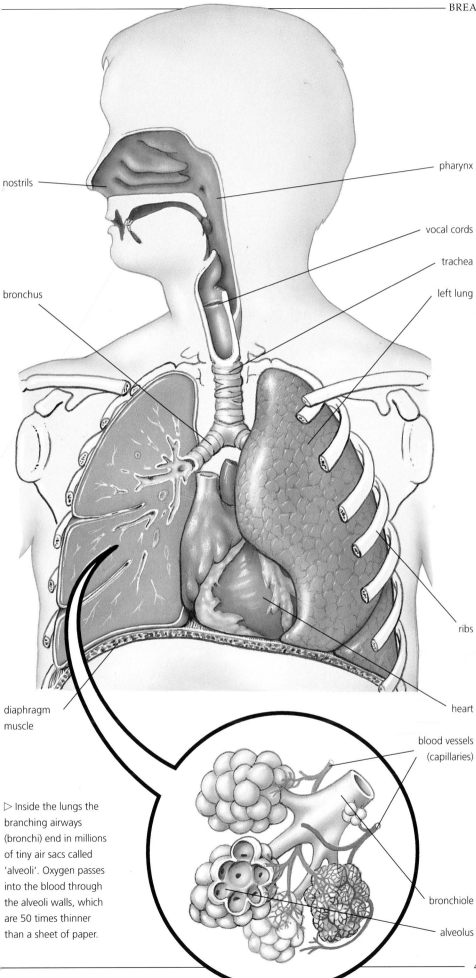

nostrils

pharynx

vocal cords

trachea

bronchus

left lung

diaphragm muscle

ribs

heart

blood vessels (capillaries)

bronchiole

alveolus

▷ Inside the lungs the branching airways (bronchi) end in millions of tiny air sacs called 'alveoli'. Oxygen passes into the blood through the alveoli walls, which are 50 times thinner than a sheet of paper.

Each bronchus supplies air to one of your lungs. In the lungs the bronchi divide again and again into a branching tree of smaller and smaller tubes called 'bronchioles'. The ends of the smallest bronchioles expand into bunches of tiny air sacs called 'alveoli'. There are more 600 million alveoli in your lungs. If you could open up all the alveoli and spread them out they would cover the floor, walls and ceiling of a small room.

This huge surface area means that the blood can soak up as much oxygen as possible from the air you breathe in. Each alveolus is surrounded by tiny blood vessels. The oxygen molecules can pass through the thin walls of alveoli into the blood vessels. They are collected by red blood cells and carried around the body.

Breathing out

The processes are reversed when you breathe out (exhale). Your diaphragm relaxes and your ribs fall to squeeze your lungs into a smaller space. This forces air out of the alveoli, through the windpipe and out of your nose and mouth. The expelled air has been changed by the respiration process. It contains less oxygen and more moisture. It also now contains quite a large amount of the gas carbon dioxide. Carbon dioxide is a waste product, produced when glucose is combined with oxygen in your cells. The same gas is produced by a burning candle flame. Carbon dioxide is transported away from your body cells by the blood. It passes out of the blood into the alveoli in your lungs, where it is breathed out.

Coughs and sneezes

When you sneeze, air is expelled from your nose at 160 km an hour – as fast as an express train! A sneeze is an automatic reaction to clear the upper airways when they have become blocked or irritated, for example by dust or pollen.

Coughing clears the lower airways in a similar way. When you are about to cough you first breathe in. Then you start to breathe out, but keep the exit to your windpipe in your throat closed. This builds up pressure which you suddenly release by letting the air out in a rush. The blast of air dislodges any obstructions.

GOOD VIBRATIONS

Nature does not miss an opportunity. Animal lungs evolved to supply oxygen for respiration. But why not also use the air flow in and out of the body to make sounds? When you let down a party balloon the narrow neck makes the escaping air vibrate, producing a sound. You can change the sound by stretching the rubber with your fingers. If the rubber is stretched more tightly it vibrates more quickly and the sound is higher.

A pair of vocal cords in your throat produce sounds in just the same way. If you place your hand on your neck you can feel the vibrations as you speak. When you sing, your vocal cords relax to produce the low notes, and they stretch tight to make higher ones.

Your vocal cords provide the sound for talking and singing, like the reeds on an oboe or saxophone. But your mouth and nasal cavities are the instrument that shapes the sound and turns it into a beautiful song, a dramatic speech or an ear-piercing scream. You change the shape of your mouth to form the musical vowel sounds (a, e, i, o, u) which you can sing as steady notes. You move your tongue and lips to add consonants (b, c, d, f, g, etc.) into the sounds from your throat, in order to form words.

You can whisper and whistle without using your vocal cords, but to really make yourself heard you must take a deep breath, open your mouth wide and shout out at the top of your voice. In shouting and screaming contests people have produced short bursts of sound of 120 decibels, almost as loud as a jet aircraft taking off. Trained opera singers can control this power to fill a theatre with beautiful sound. If you play a wind instrument you also have to learn to control your breathing when playing.

▽ The narrow gap between your vocal cords is the source of most of the sounds you make. After puberty male vocal cords are longer and vibrate more slowly than female ones, so men have lower-pitched voices. The lowest notes sung by male bass singers vibrate about 75 times each second. A female soprano can make her vocal cords vibrate more than 1000 times a second.

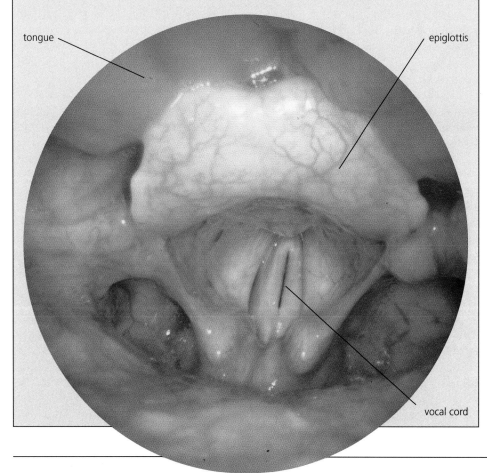

tongue

epiglottis

vocal cord

When someone else has the hiccups it can be amusing, but when you have them it is irritating and uncomfortable. Imagine not being able to stop. Charles Osborne of Iowa, USA, hiccuped for 68 years! When you hiccup your diaphragm jerks suddenly, forcing a short blast of air from your lungs. Hiccups usually follow a meal, especially when you have eaten too much or too quickly. Your overfull stomach presses against your diaphragm, irritating it and making it move unpredictably. As the food is digested and the stomach contracts, the hiccups subside. Sometimes you can cure hiccups by holding your breath or sipping water from a glass. The pause in your breathing gives your diaphragm a chance to settle back to its normal rhythm. A sudden shock or surprise, which makes you stop breathing for a moment, can do the trick as well.

Breathing and exercise

Your normal breathing rate is about 10 to 12 breaths per minute. When you are relaxed you breathe quite gently. At each breath you inhale and exhale about half a litre of air, about two cupfuls. This is only about one-sixth of the total capacity of your lungs. When you start exercising you breathe more heavily. If you are working hard but steadily, for example in a workout to music, you breathe more often and more deeply to supply the extra oxygen your muscles need. This is called 'aerobic' respiration and gives its name to the exercise called aerobics. Steady aerobic exercise is the best way to increase the fitness of your heart and lungs.

If you start working even harder, for example in the sprint to the finish of a race, then you soon start gasping for breath. Now your heart and lungs cannot keep up with oxygen demands from your muscles. The extra energy is supplied by the muscle cells, which burn up their glycogen 'anaerobically'. This means that energy is extracted from the glycogen without using oxygen. This is an inefficient process that produces a waste product called lactic acid. The lactic acid and lack of oxygen make your muscles ache and you feel out of breath. After the final burst for the tape, athletes continue to breathe heavily for several minutes. They need extra oxygen to restore normal aerobic respiration.

blocked with sticky mucus. This makes it more difficult to breathe. A doctor checks with a peak flow measurement to see if a patient's lungs are affected. The patient blows as hard as they can to move a pointer along a scale. If their airways are narrow they cannot blow as hard as when they are clear. An asthma attack can be relieved by breathing in a drug from an inhaler that relaxes the muscles in airway walls, allowing them to open up.

In many cases asthma seems to be a type of allergy. An allergy occurs when your body's defence systems overreact to an everyday substance in the environment. Many people are allergic to grass pollen, for example, and suffer from hay fever during the late spring and early summer. Asthma attacks are known to be triggered by a number of factors including stress, and dust and pollution from factories and motor vehicles.

Allergies and asthma run in families. The tendency to develop these problems is probably passed on from parents to children through the genes. But not every child in a family develops asthma. It seems that something may make a particular child's airways sensitive, making them likely to have asthma attacks later in life. Recent research has shown that children are more likely to develop asthma if they are exposed to the droppings of house-dust mites (tiny animals that live in the dust in all our homes) or cigarette smoke, before the age of one. Perhaps it is the pollution we trap in our warm, poorly ventilated homes that is responsible for the current increase in asthma attacks.

Short of breath

Gasping for breath at the end of a race is nothing to worry about, but about one in ten people experience spells when they feel out of breath even when going about their everyday life. These people are suffering from asthma. Asthma is one of the few diseases that is becoming more common in the western world. In the United Kingdom it is now the most frequent cause for children being away from school.

During an asthma attack the linings of the airways become irritated or inflamed, and the airways become narrow and

△ Mountaineers struggle for breath in the thin air on top of a mountain. They gradually adapt by producing more red blood cells to absorb oxygen.

▷ Breathing in dust can damage the lungs and trigger breathing problems. Farmer's lung is a lung disease caused by dust from mouldy crops.

NERVES AND SENSES

nerve fibre

Imagine floating in the dark without sounds, sights or smells. You feel neither hot nor cold, heavy nor light. Without your senses you would be isolated from the world. You could not feel the pleasure of a cool breeze on a summer morning, nor the pain that makes you let go of a hot saucepan to save your skin from burning.

To stay alive your body needs a constant stream of information about its surroundings. You must be able to sense danger and be ready to respond in an instant. Sensing and responding are the jobs of the nervous system.

A bundle of nerves

A delicious smell, a suspicious sound or a sudden change in temperature create streams of electrical impulses that run through your nervous system. This system is built from specially adapted cells called 'neurons'. Bundles of neurons form nerve fibres. A thick bundle of nerve fibres emerges from the base of your brain, forming your spinal cord. The brain and the spinal cord together make up your 'central nervous system'. This branches into a network of nerves that extends into every part of your body, carrying signals from your eyes, ears and other sense organs to your brain, and sending instructions to your muscles.

Two types of nerve cells carry information around your body. 'Sensory' neurons carry messages from your body to the spinal column and then on to the brain. 'Motor' neurons carry signals from the brain or spinal column to your muscles, to stimulate them to contract. The ends of sensory neurons are adapted to respond to different kinds of signal such as temperature, pressure or vibration. All the signals that are carried by the nervous system must be organized and this is the brain's responsibility – it is your body's control centre.

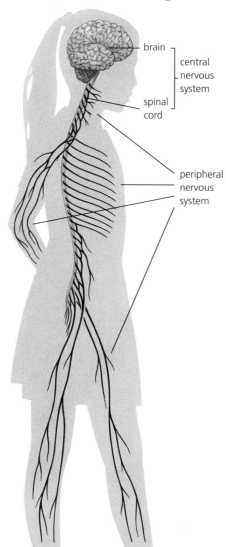

brain

central nervous system

spinal cord

peripheral nervous system

△ Spinal nerves extend into every part of your body. There are also 'cranial nerves' (not shown), which connect directly into your brain from within your head, for example the optic nerves.

Inside your brain billions of specialized neurons process the signals from your senses and make you aware of the world outside your body. But how do electrical pulses in your brain create the sensation of listening to a beautiful piece of music? We understand how individual neurons carry signals from our sense organs to our brains and how signals from the brain make our muscles respond. But exactly how these signals create our thoughts and feelings is still a mystery.

Live wires

Your nervous system is like a large communications network. Each neuron can transmit and receive signals from many other nerve cells. The body of a typical neuron extends into a long fibre called an 'axon' and a series of twig-like branches called 'dendrites'. Axons are the nervous system's wires. A single axon can be up to a metre long, for example those in the nerve which runs from your spine to your big toe. The end of an axon breaks up into many fine branches, tipped with tiny bulb-shaped knobs. These nerve endings connect to the dendrites and bodies of other neurons through junctions called 'synapses'. There are more wires and connections in your nervous system than in the telephone network worldwide.

A neuron transmits signals by a special mechanism unlike anything found in the non-living world. The signal travels as a pulse – rather like the 'Mexican wave' at a sports stadium. Because each person taking part in a Mexican wave stands and then sits a fraction of a second after their neighbour, a pulse appears to travel through the stand. A nervous impulse travels along a neuron in a similar way, as charged atoms move in and out of the axon through tiny channels in its walls, just like the spectators standing and sitting in their seats. When these electrical pulses reach synapses they trigger the release of tiny

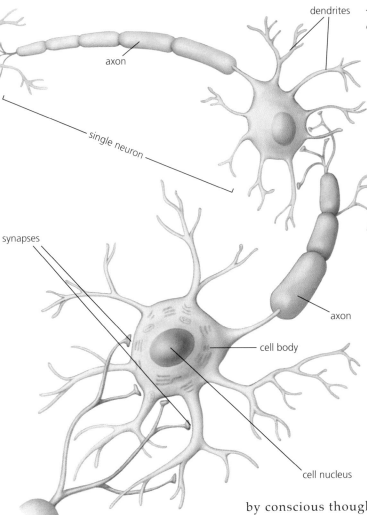

axon

single neuron

dendrites

◁ The nerve fibres that extend throughout your body are bundles of individual nerve cells or neurons. Each neuron links to many others through gaps called synapses. Chemical pulses carry nerve signals from one neuron to another across these gaps.

synapses

axon

cell body

cell nucleus

quantities of chemicals called 'neurotransmitters'. These chemicals carry nerve signals across the narrow gaps that separate one nerve cell from another. A single neuron may receive chemical signals from many other neurons that connect to it. The combined effect of these signals determines the rate at which the receiving neuron produces pulses, or 'fires'.

In control

Some of the actions of your nervous system are under your conscious control. For example, if you decide to put down this book and get a drink, then your movements result from your decision. This type of action is described as 'voluntary'. However, many of the signals carried by your nerves are involuntary. You do not have to decide to make your heart beat, nor to make your muscles shiver when you are cold, and you cannot stop these processes

by conscious thought. The nervous signals that stimulate these automatic processes are described as 'autonomic'.

Reflexes and reactions

Pulling your hand back from a sharp point or a hot plate is an example of a 'reflex'. You do not have to decide to move, you just do so automatically. This type of reflex is controlled by a simple nerve circuit called a 'reflex arc'. Sensory neurons detect the sharp pressure or the heat, and generate impulses which travel to the spinal column. Here the sensory neurons connect with motor neurons, which send signals to make the appropriate muscles in your arm contract. There are also links to neurons which connect to your brain, so you are aware of the pain – but not until after you have made the response, since the signal has further to travel to get to your brain than to pass around the arc.

You can sometimes suppress a reflex by an effort of will. If you decide to keep your grip on an object which is hot, or to prick your finger until it bleeds, then the strength of the nerve impulses from your brain outweighs those from the reflex arc. Some people who practise the Indian art of yoga are able to exercise considerable control over pain.

It takes about one fiftieth of a second for a nervous signal to travel from your eye to your brain and then to your hand. This is your fastest possible reaction time. Sports players practise reacting as rapidly as possible to high-speed events, such as saving a penalty kick or returning a powerful serve on the tennis court.

△ Sometimes normal reactions are not fast enough. A top baseball player must anticipate the ball's flight to make a hit.

Five senses?

Traditionally we have five senses: sight, hearing, touch, taste and smell. But we have many more than five kinds of sensation. We sense both the colour and the brightness of light with our eyes. Our ears can detect both the loudness and the pitch of a sound. They also contain sensors that help us to balance by detecting changes of movement in our bodies. The sensations of taste and smell are produced by chemical sensors on our tongues and in our noses, which can distinguish thousands of chemical compounds, some in minute quantities. Touch is a combination of sensations produced by pressure and temperature. Our muscles have stretch sensors that detect their position. For example, can you tell if your legs are straight or bent without looking?

If we lose one of our senses, our other senses may be intensified to make up for it. For example, people who have lost their sight develop a very good sense of hearing and touch.

▽ In daily life our senses are bombarded with an incredible mixture of information. Our brain allows us to notice the smell of a particular fruit, the sound of a certain voice or the feel of a piece of cloth while keeping the other sensations in the background.

Seeing is believing

Each of your eyes is a delicate mechanism for collecting and concentrating light to make a colour image of the outside world. Light enters the eye and is focused on the 'retina' by a lens, similar to the way a camera lens focuses light on to a film. The eye lens is flexible and its thickness can be adjusted by the 'ciliary muscles' to focus on objects at different distances. If the adjustment of your eye lens is not well-matched to the length of your eyeball, you will either be short or long-sighted. If you can see close-up objects clearly but see objects at a distance as a blur, you are short-sighted. If your vision is blurred when objects are close, but you can see things quite clearly at a distance, you are long-sighted. The blurred image produced can be brought into focus with the aid of spectacles or contact lenses.

The 'iris' controls the amount of light entering the eye, like the aperture in a camera. In low light the iris opens and the pupil is large. In bright light the iris closes and the pupil shrinks. The muscles of the iris are controlled by an autonomic reflex. Stand in front of a mirror and try to change the size of the iris by concentrating hard – you cannot. But if you shine a light into your eye, the iris adjusts automatically.

In the retina there are special light-sensitive cells (receptor cells) which contain chemicals that absorb light energy. There are two distinct types of receptor cell, called rods and cones because of their shapes. The rods are sensitive to very weak light, but they do not detect its colour. The cones are colour-sensitive, but they require more light than the rods to make them active. This is why we cannot see colour well by the light of the Moon. The Moon's weak, reflected light is only sufficient to activate the rods. In bright sunlight, however, the cones are fully-activated and we can see vivid colours.

Our eyes are incredibly sensitive to colour. We can probably detect more than 10 million different shades. But the retina contains just three different kinds of light-

sensitive cone cell, each of which is stimulated most strongly by a particular 'primary' colour – red, green or blue. So how do we see colours like yellow or orange?

Yellow light stimulates both the red and the green receptor cells in the retina, and the brain interprets the combined signals from these cells as yellow. Orange light stimulates the red receptors a little more strongly, and the green receptors a little less strongly, than yellow light. We can thus detect many more than three colours by using the three types of receptor cell in combination. But we can be fooled. A mixture of pure red and pure green light stimulates the red and green receptors in the same way as pure yellow light, so we see this mixture as yellow too.

By combining the three primary colours in different proportions we can create all the colours the human eye can detect. It is because of the way our eyes work that there are just three primary colours. It is amazing to think that all the pictures on our television screens are made up of these colours only. If the human eye had five kinds of light-sensitive cell, then five primary colours would be needed.

▷ Whose eyes are these? Eyes are as individual as fingerprints. We look into someone's eyes to try to guess their thoughts and feelings.

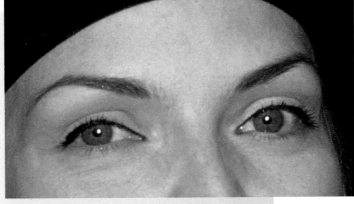

ANSWER: Michelle Pfeiffer

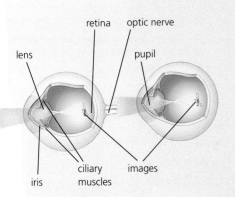

lens · retina · optic nerve · pupil · iris · ciliary muscles · images

△ The images formed by the eye lens on the retina are actually upside down. But 'seeing' does not take place in the eyes. Our brains process the signals from our eyes to produce the view of the world we see in our minds. This is the correct way up.

COLOUR-BLINDNESS

Some people are colour-blind. This does not mean that they see everything in black and white, but that they have difficulty in distinguishing the same number of colours as people with normal colour vision. The usual cause is that the green-sensitive cone cells do not function properly, so the colour-blind person is less sensitive to the difference between red and green light. Red–green colour blindness is often inherited and is carried by a gene on the X chromosome (see page 68). About 1 in 12 men are red–green colour-blind, compared to 1 in 250 women.

Sounds good

Our sense of hearing is usually thought of as second in importance to our sense of sight, and it is true that we obtain most of our information about the world through our eyes. But in some ways our ears are more sensitive than our eyes. When coloured paints or lights are combined our eyes see the mixture as a single colour. But when several sounds are mixed together, we can still pick out each sound individually. For example, when you listen to an orchestra you can pick out the high-pitched flute and the low-pitched double-bass as they play together.

Sounds are vibrations of the air. Sound waves travel down the ear canal and hit the eardrum, a tightly-stretched piece of skin, making it vibrate. This movement is transmitted through the ear to the fluid inside a spiral tube called the 'cochlea'. Tiny hair-like cells in the lining of the cochlea respond to the motion of the fluid by sending nerve signals to the brain.

We can hear a vast range of sounds, from deep rumbles when the air is vibrating just 20 times each second, to high-pitched whistles when it is vibrating a thousand times faster. The quietest sounds we can detect, such as a falling leaf, have more than a billion times less energy than the loudest sounds we can stand without feeling pain. Very loud sounds, such as explosions, jet aircraft taking off, or even a personal stereo turned up too high, can damage your hearing.

A BALANCING ACT

Your ears help to maintain your balance by detecting the slightest movement of your body. Sitting on top of the cochlea are three bony tubes called the 'semicircular canals'. These contain fluid which responds to movements of your head, like coffee sliding around in a cup on a moving train. The moving fluid bends flexible stalks linked to nerve cells inside each of the canals, sending signals to your brain. To stay balanced you need three canals to detect three types of motion – nodding, twisting and side-to-side bending. When you are spun on a roundabout, the fluid in the semicircular canals keeps moving for a few seconds after you stop, and this makes you feel dizzy.

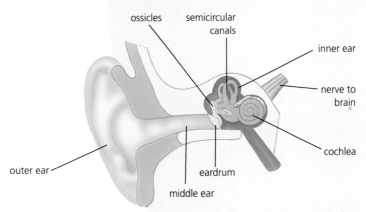

ossicles · semicircular canals · inner ear · nerve to brain · outer ear · eardrum · middle ear · cochlea

△ Your ears contain the smallest bones in your body. These are the three 'ossicles' which amplify the tiny vibrations of the eardrum and transmit them to the fluid inside the cochlea.

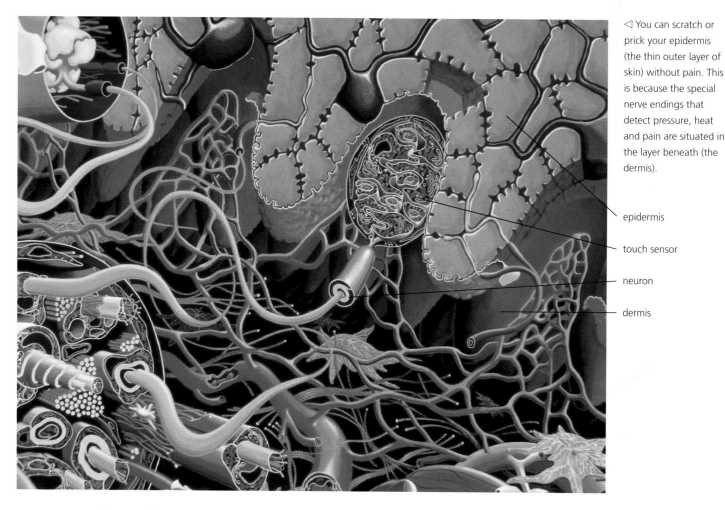

◁ You can scratch or prick your epidermis (the thin outer layer of skin) without pain. This is because the special nerve endings that detect pressure, heat and pain are situated in the layer beneath (the dermis).

epidermis

touch sensor

neuron

dermis

In touch

When you touch something, or something touches you, you can feel various sensations including pressure, pain, tickle, itchiness, heat or cold. These feelings are created by signals from a variety of specially adapted nerve endings in your skin. Disc and bulb-shaped sensors at the nerve tips convert various kinds of force applied to your skin into nerve signals. Some of these sensors are sensitive to light pressure, for example a brush from a feather; some respond to stronger pressure such as someone gripping your arm; some detect vibrations and create the tickling sensation that can make you laugh uncontrollably. Pain and temperature are sensed by 'free' nerve endings, which are nerve ends that do not have a special structure at their tips.

The numbers of nerve endings in your skin varies greatly over the body's surface. The fingertips, palms, lips and tongue have the highest density of touch sensors. We use these body parts to feel the shape of objects and test the texture of food.

There are more than 200 touch-sensitive nerve endings per square centimetre of skin in your fingertips. With practice a blind person can read the raised dots of braille letters as quickly as a sighted person can read print.

There are far fewer nerve endings in the skin on your back, legs and arms. This explains why it is sometimes difficult to find the exact location of an itch on your back – you must get someone to scratch over a large area to relieve the irritation. The cause of the itching sensation is not well understood. Unlike pain, which we can feel inside our body (for example in a tooth) as well as on the outside, itching only seems to occur in the skin and eyes. Certain chemicals, including some made by the body itself when it is injured, cause the skin to itch.

Pain, the most unpleasant of all sensations, occurs when body parts containing pain-sensitive nerve endings are being damaged. It warns us that something is wrong, and signals us to take steps to protect our bodies. Unfortunately, once the damage is done the pain often continues. It is even possible to feel pain in a limb that has been severed in an accident. The nerves that once carried signals from the limb continue to function and the victim feels sensations from a 'phantom' limb that no longer exists.

Smell

Human beings can distinguish between 10,000 and 40,000 different smells. Inside your nose there are 'smell' sensors which respond to chemicals in the air breathed through your nose. The chemicals make the nerves send messages to your brain, which produces your sense of smell. With practice we can pick out many different smells at the same time. For example, in a room you might smell a mixture of food, polish, perfume and coffee. Wine tasters learn to describe the smell of a wine in terms of a mixture of basic smells with which they become familiar, such as flower scents, fruits, earth, and so on.

'How can you work in that smell?' After a while someone who works with a strong smell, such as a pig farmer or a zoo keeper, ceases to notice it. Their sense of smell becomes habituated. Habituation occurs with our other senses too. The loud ticking of a new clock can be irritating at first, but we soon learn to ignore it. When you first get into a cold swimming-pool or a hot bath you may feel uncomfortable. After a while your nerves become habituated to the stimulus, and you adjust to the temperature. But the ability of our senses to adjust is limited. We would eventually jump out of the bath if the temperature started to rise.

▷ A fragrance tester spends many years developing a 'nose'. He learns to distinguish thousands of different odours.

A budding taste

The sensation of taste is created by about 100,000 taste buds you have on your tongue and on the inside of your mouth and throat. These contain nerves which transmit information about taste to the brain.

Our taste buds are sensitive to just four basic flavours – sweet, salty, sour and bitter. Sweetness and salt produce familiar sensations. Sour is the effect produced by acids such as lemon juice or vinegar. Chemicals called alkalis, including soap, and substances found in radishes and coffee beans, produce a bitter taste. Our enjoyment for the taste of food is probably actually more to do with its smell than its taste. When you have a bad cold the taste buds in your mouth are still working normally, and you can tell a salty drink from a sweet one. But most of the flavour of your food will be gone. This is because the smell sensors in your nose are blocked. The full taste sensation created by food is not just generated by the taste buds. The feel of the food in the mouth, its smell, its appearance, its familiarity, and even your feelings about a particular type of food, can all affect your enjoyment.

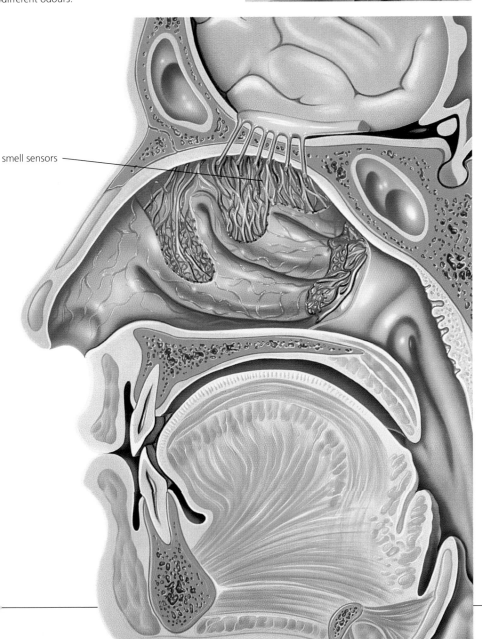

smell sensors

◁ The nerve endings that detect smell are located on a bulb high in the nasal cavity. Sniffing increases the air flow over the 'olfactory bulb', bringing more molecules into contact with the smell sensors.

▽ Your taste buds are on the surface of your tongue. Food molecules enter the taste buds through tiny pores and stimulate special nerve cells inside.

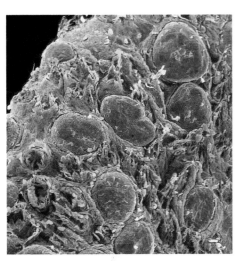

HORMONES: CHEMICAL MESSENGERS

How does your body know how big to grow, how much fuel to burn in its cells, or when to mature from childhood to adulthood? Body processes like these are controlled by an army of chemical messengers called hormones.

Throughout your body there are numerous glands. These are groups of specialized cells which act as chemical factories to produce some of the special chemicals your body requires. Some glands deliver their products through tubes called ducts directly to where they are needed. For example, the sebaceous glands release sebum to the skin, making it waterproof. The salivary glands release saliva into the salivary ducts, which carry it to the mouth. Other glands, however, do not have ducts. These are the endocrine glands, which produce chemicals called hormones. Hormones are released directly into the blood, which carries them to the parts of the body where they are required.

Hormones are chemical messengers. Cells respond to their presence in the blood by switching some of their activities on or off. For example, a hormone called prolactin, produced by women who are breastfeeding, stimulates special cells in their breasts to make milk. There are dozens of different kinds of hormone, each with a different job to do. Each hormone is a molecule with a specific shape, which only certain cells in the body will recognize. When the hormone reaches its target cells it sticks to their surface, or passes into them through the cell wall, to stimulate their activity. The more hormone there is present, the more active the cells become.

▷ How are you feeling? Perhaps you are hungry, sleepy, energetic or nervous? Many of our vital processes are controlled by hormones – chemical messengers made by numerous glands throughout our bodies.

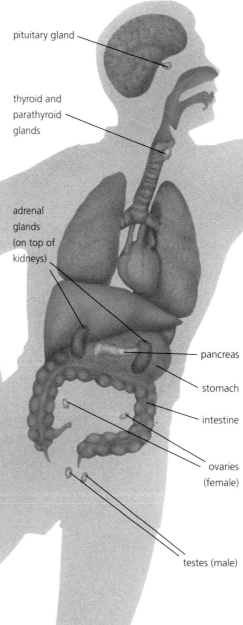

pituitary gland

thyroid and parathyroid glands

adrenal glands (on top of kidneys)

pancreas

stomach

intestine

ovaries (female)

testes (male)

All sorts of gland

Your main hormone-producing glands are the pituitary gland just below your brain, the thyroid and parathyroid glands in your neck, the adrenal glands on top of your kidneys and the sex glands at the base of your trunk (in females these are the ovaries which are internal; in males they are the testes which are external). Hormones are also produced by other organs including the pancreas, the thymus and the lining of the intestine.

Hormones from the thyroid control your metabolic rate – the rate at which your body cells burn fuel. If your thyroid is not active enough you become lethargic and put on weight. If it is working too hard you become overactive and anxious, you eat more but still lose weight, and your eyes may start to bulge. The thyroid gland needs iodine to make its hormones. If a child's diet lacks iodine early in its life the child will fail to grow properly and may suffer permanent brain damage. This condition is known as 'cretinism'.

A hormone from the parathyroid gland controls the amount of calcium in the blood. It acts on bone cells and cells in the intestine to make them release or absorb calcium.

Fight or flight?

Your adrenal glands produce the hormones adrenalin and noradrenalin in response to a stressful situation such as waiting for the starter's gun in a race or taking an exam. These hormones have several effects designed to help you cope with danger. They make your heart beat faster, speed up your breathing and increase the rate at which your muscle cells burn fuel. When you feel 'butterflies in your stomach' and turn pale, this is due to the lack of blood in your skin and digestive system. As much blood as possible is diverted to your brain and muscles where it is needed in the emergency.

The adrenal glands also produce a number of hormones known as steroid hormones. The most important of these are aldosterone and cortisol. Aldosterone helps control the amount of fluid in your blood. When you have not had a drink for some time aldosterone is produced to signal to your kidneys to conserve water.

◁ The sight and sound of a fierce dog stimulates your body to release the hormone adrenalin. Your heart pounds and your muscles tense, ready for a quick escape!

the body put on weight, increasing the amount of muscle. They were originally developed to help victims of starvation and cancer patients. Now, however, these drugs are sometimes abused by athletes and body builders. Athletes undergo drug tests to check that they are not improving their performance with such drugs. Those who 'test positive' are banned from competition. The use of anabolic steroids can produce severe side effects, liver cancer for example, when used in this way for long periods.

The thymus is an organ located beneath your breast bone. It manufactures special white blood cells called 'lymphocytes', which are part of the body's defence

MYSTERY GLAND

The pineal body is a tiny gland not much bigger than a pin head, located in the brain. Its function is still rather a mystery. In simple animals, including primitive fish, it acts almost as a third eye. In humans it is not light sensitive itself, but it is linked to the eyes via the brain. When it gets dark the pineal body produces a hormone called 'melatonin'. Patients injected with melatonin feel sleepy, and it may be that this hormone helps to control body rhythms such as sleep. Young children produce high levels of melatonin and sleep far more than adults. It is also thought that melatonin prevents children's bodies from maturing too early. The amount of melatonin produced by the pineal body decreases during adolescence.

Cortisol helps control the rate at which body cells make proteins. It also stimulates the conversion of amino acids (see page 36) into glucose. These actions help the body to cope with starvation or stress.

Growing up

Steroid hormones released by the sex glands at puberty trigger the development of the reproductive organs and the growth of pubic hair. In females these hormones are called 'oestrogens'. The main male sex hormone is called 'testosterone'. These hormones also stimulate the development of the 'secondary sexual characteristics', which distinguish men from women. Male voices deepen, their beards start to grow and their muscles increase in size. Female bodies become more rounded, breasts start to develop and their periods begin.

The regular release of an egg from the ovaries in women is controlled by hormones from the pituitary gland. Oestrogen hormones and a hormone called 'progesterone' are produced by the ovaries during the monthly cycle. They stimulate the lining of the womb to thicken

in preparation for pregnancy. If the egg is not fertilized (see page 67) then the hormone levels fall and the womb lining is shed along with some blood. If the egg is fertilized, progesterone production continues to keep the womb in the right condition for the fertilized egg to develop. The placenta, which links the mother to her growing baby, also contributes to the production of the necessary hormones.

Anabolic steroids are drugs which have similar effects to testosterone. They make

▷ Boy sopranos have not yet reached puberty. Within a few years testosterone will make their voices break. Then they will join the tenors and basses.

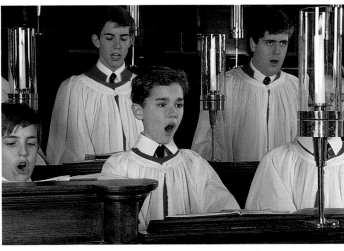

INSULIN'S STORY

The pancreas, located between your stomach and intestines, is an important part of the digestive system, producing enzymes needed to break down food (see page 39). It also produces two important hormones called insulin and glucagon. These hormones work together to control the amount of sugar in the blood.

After you have eaten a meal your blood sugar (glucose) level rises. This stimulates the pancreas to release insulin. Insulin helps glucose molecules move through cell membranes into cells, where it can be used as fuel. Insulin was discovered in 1921 by Canadians Frederick Banting and Charles Best and Scotsman John Macleod.

If you have been working hard and have not eaten for a while your glucose level falls. This stimulates the pancreas to release the hormone glucagon. Glucagon stimulates cells in the liver to convert stored glycogen (see page 40) into glucose to raise your blood sugar level.

Between them insulin and glucagon perform a balancing act to maintain the correct level of glucose in your bloodstream. But things can go wrong. If the pancreas does not produce insulin then cells cannot absorb glucose from the blood. The person becomes very weak. Fat and protein stored in the body are used to provide energy instead, but this produces waste products which are harmful to the body. The patient may enter a coma (become unconscious). These are the symptoms of the disease diabetes.

Diabetes is a common condition, and it can be fatal if it is not treated. The usual cause of diabetes in young people is the failure of the pancreas to produce insulin. When diabetes occurs in middle age, however, it is often associated with obesity (being severely overweight). The body does not produce enough insulin to cope with the high sugar intake. There are now more than 10 million diabetics in the United States alone and, as western lifestyles spread, numbers are increasing worldwide.

◁ Diabetes can be treated with regular insulin injections. Some diabetics now carry miniature pumps that feed insulin into their bloodstream steadily throughout the day.

system against invading germs (see page 43). The thymus produces hormones which control the production and activity of these cells. The thymus is the biggest gland in a newborn baby, who must develop resistance to many common infections. But as we grow up and our immunity develops, the thymus becomes relatively smaller and less active.

Keeping control

The endocrine glands around the body are rather like the players in an orchestra. Each one has a job to do, but they must all work together to produce a balanced result. Conducting your body's orchestra of glands is the job of the pituitary gland and the hypothalamus, the region of the brain just above the pituitary gland.

How can the sight of a fierce dog make you turn pale, or the smell of food cooking make your stomach grumble? When you confront that fierce dog your senses detect its threatening appearance and barks. Nerve signals are transmitted to your brain from your eyes and ears. The role of the hypothalamus is to translate this nervous activity into chemical signals. The hypothalamus releases hormones which stimulate the pituitary gland. The pituitary gland then releases more hormones which stimulate the activities of the other glands in your body appropriately. A chain of chemical messages, carried by your blood through your body, translates your thoughts and feelings into chemical action.

As well as controlling the other endocrine glands, your pituitary gland produces hormones that cause long-term changes to your body. Growth hormone

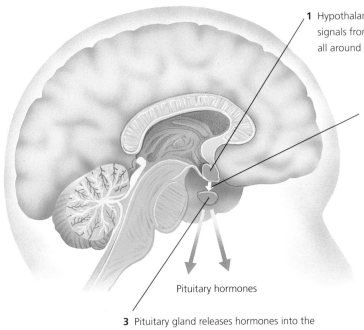

1 Hypothalamus receives nerve signals from the brain and sensors all around the body

2 Messages are passed on by hormones to the pituitary gland

Pituitary hormones

3 Pituitary gland releases hormones into the bloodstream to stimulate other glands into action

△ The hypothalamus and the pituitary gland work together as a control centre, monitoring the levels of different hormones in the blood and providing a link between your nervous system and your endocrine (hormone) system.

controls your growth between childhood and adulthood. The pituitary gland releases it into your bloodstream during the night. If it fails to produce sufficient growth hormone children may be small for their age and may not reach full adult height. Fortunately, these children can now be treated with growth hormone to make sure that they develop normally.

If the pituitary produces too much growth hormone the result is a condition known as gigantism. People with gigantism usually have a shorter life span than normal because they often develop other disorders and easily pick up infections.

Hormones and emotions

Strong feelings and emotions are associated with the release of certain hormones into the blood. Hormones are involved in our feelings of anger, excitement, fear and even sexual love. At puberty, when the sex glands start to be active, young people experience a wide range of new feelings about themselves and their relationships with others.

Do hormones cause these feelings and emotions? Or do our thoughts and feelings cause the release of hormones, to prepare our bodies for action? This is probably a question similar to the famous 'Which came first – the chicken or the egg?', the answer to which is neither – you cannot have one without the other.

▷ A faulty pituitary gland can produce exceptional growth (gigantism). This man is about 2.5 m tall, more than half a metre taller than an average person.

THE BODY IN BALANCE

Whether you are baking in the sun or freezing in a blizzard your body temperature should stay close to 37°C. Your body automatically adjusts to changes in the outside world, to maintain a constant environment beneath your skin.

In an air-conditioned building the temperature and humidity (the amount of moisture in the air) are the same all year round. If it is snowing outside, heaters warm the air. If there is a heatwave, refrigeration units cool it down. If the air is too dry or too damp, machines called humidifiers adjust its moisture content. All these adjustments are made automatically. Thermometers and moisture sensors are constantly keeping check on the conditions inside the building, and passing the information back to a control system. If conditions change from those required, the system adjusts the air-conditioning appropriately.

The conditions inside your body are maintained in a similar way. In order for your body's processes to operate efficiently its internal temperature must be about the same in all weathers and the concentrations of salt, glucose and other chemicals in your blood must not change too much, even though you only eat now and again. This steady condition of the inside of your body is called 'homeostasis'.

Your body maintains homeostasis by using various sensors that detect any changes taking place. Sensors in your skin and brain check the inside and outside temperature of your body. Special sensors, for example, in a part of your brain called

the hypothalamus, check the amounts of different chemicals in your blood. If something needs adjusting then signals are sent out to start the changes that will bring your body back into balance. These signals are produced by your body's two control systems: the nervous system, which is under the direct control of your brain, and the endocrine (hormone) system (see pages 58 to 61).

Hot and cold

Your body has several mechanisms for losing heat if it becomes too hot, or retaining heat when it starts to cool down. Some, such as sweating and shivering, are automatic. You can control others yourself, by putting on a warm jumper or moving out of the hot sun, for example. These responses allow you to survive in a greater range of temperatures than any other creature, from the frozen Antarctic wastes to the hot desert sands.

Feeling cold

You feel chilly as soon as the temperature of your skin starts to fall. Your body's control systems respond to this 'cool' signal by narrowing the blood vessels just under your skin's surface so there is less blood flowing in them. This explains why

◁ Your body is not completely at the mercy of the elements. It is equipped with feedback systems that keep it warm in the cold and cool in the heat. Temperature sensors in your skin and internal organs produce signals to make you shiver or sweat.

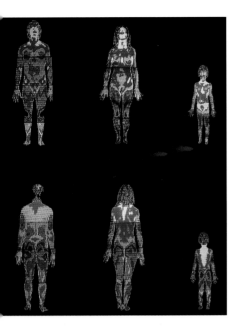

△ Variations in skin temperature over the human body. White areas are hottest, purple areas coolest.

Your hands and face become paler in the cold. Usually your blood carries heat from the centre of your body to your hands and feet, where it is lost through the skin; so reducing the blood flow to the skin helps to reduce heat loss from your body.

If your body continues to lose heat, temperature sensors in your brain sense that your internal blood temperature is starting to fall. Then you start to shiver. The shivering movements of your muscles generate heat that helps to maintain your internal body temperature. At the same time, your brain signals to you to put on more warm clothing, become more active or eat and drink something hot.

If you go boating in cold weather or walk in the mountains in the winter, it is important to recognize the signs of hypothermia. Hypothermia occurs when the body is unable to maintain its internal temperature. Falling into cold water or exhaustion in cold conditions can lead to rapid loss of heat that the body's internal systems cannot replace. As a hypothermia victim's temperature falls to 35°C they become sluggish and confused, and may start to shiver uncontrollably. Body movements become uncoordinated, speech is slurred and the memory vague. In these circumstances you must stop, find shelter, remove the wet clothes, put the person into a warm dry sleeping bag, and give them food and a hot drink.

SHIPWRECK

If you are unlucky enough to be shipwrecked in cold water, hypothermia is your greatest enemy. Water carries heat away from the body much more rapidly than air. If you have time before the boat sinks, you should put on as many clothes as possible to retain heat. If you have a lot of body fat (which provides insulation and fuel to give you energy) you will probably survive longer than a thin person. Whatever your build, after a few hours in warm sea water, or just a few minutes in freezing water, your internal temperature will start to drop and the symptoms of hypothermia will appear. As your body temperature falls below 33°C the body's mechanisms for keeping itself warm will fail, and your temperature will start to fall more rapidly. When it drops below 30°C you will lose consciousness. The lowest internal temperature which your body can survive is about 25°C.

Stay cool!

If you start to feel hot, because of vigorous exercise, a hot environment or both, your body's cooling mechanisms become active. The blood vessels under the skin dilate (expand) to increase the blood flow to the surface, allowing extra heat to escape. Because more blood is close to the surface of your skin, your skin looks redder. Then you start to sweat. Sweat is a mixture of salt and water. The evaporation of sweat on the surface of your skin has a powerful cooling effect. Aftershave and perfume feel cool on the skin for the same reason – they take heat away as they evaporate.

When you are too hot your brain signals that you should slow down if you are working hard, remove clothing or move to a cooler place. The competitive instinct may lead runners in a marathon race to ignore signals that their bodies are overheating. Then they are at risk of hyperthermia (increased body temperature) and heat exhaustion. During the race the runners may lose so much water through sweating that they become dehydrated. As they lose water, their blood volume decreases, their blood pressure falls and they may collapse.

◁ Heat exhaustion at the end of a long race is due to dehydration. This runner will soon recover when he stops and has a drink to replace his lost body fluids.

If you have ever stayed out in hot sunshine for too long, especially without a hat, you may have suffered from heat stroke. You have a headache, you feel feverish, your skin is hot and dry and, in severe cases, you may collapse. This is hyperthermia resulting from the heat that your body has absorbed from the sun's rays and the damage that has been done to your skin.

A fever is also a form of hyperthermia, produced when your body is fighting an infection. Some scientists have suggested that the increased temperature may help to kill off the invading microbes, or 'bugs', but this has not yet been proved.

Body fluids

If you get up and drink two large glasses of water, then sit down and start reading again, before very long you will need to go to the toilet to urinate. What has happened to the water you drank? It feels as if it has passed almost directly through your body, from one end to the other, picking up a little colour and smell on the way. But actually it has made a much more complicated journey than this.

Your body must balance the amount of water it contains, in order to keep the make-up and pressure of your body fluids within safe limits. During the day you lose water from your body in urine, faeces, sweat and when you breathe out. Typically you might lose 1.5 litres in urine and 1 litre through other processes. About half of this water is replaced from drinks and the other half from the water in your food. If you are taking part in sports that make you sweat, or suffering from an illness that causes you to lose large amounts of fluid through sickness and diarrhoea, you will need to take in a lot more fluid to avoid becoming dehydrated.

Most of the water from drinks and food is absorbed into the blood through the intestine walls during digestion (see pages 39 to 40). A small quantity remains in the faeces to make it easier for them to pass out of your body. Water entering the blood increases the blood volume, raising its pressure and stretching the walls of blood vessels. The amount of water in blood is sensed by nerve endings in certain blood vessel walls, which respond to being stretched. Special sensors in your brain also monitor the level of salts in the blood.

Waterworks

As your blood circulates around your body it passes through your kidneys. The two kidneys are filters which clean the blood and extract the water that your body does not need. Urine consists of waste materials from the blood dissolved in water. It drains from the kidneys into your bladder, a stretchy muscular bag which can expand to hold a considerable volume of urine. The bladder of most people has a capacity of about half a litre, but in some the bladder stretches to hold almost one litre of urine. The frequency with which you need to go to the toilet depends on both the size of your bladder and the amount you drink.

Your hypothalamus uses the signals from the various sensors around your body to control your water input and output. The walls of the bladder contain stretch sensors like the walls of some blood vessels. As soon as your bladder contains more than about one third of a litre of urine you want to urinate. If you have not had a drink for a while, your body's water levels start to fall. The hypothalamus senses this drop, and it releases a hormone that

▷ Infant diarrhoea can be life threatening. Children die because their bodies become dehydrated. A simple rehydration therapy consists of water, sugar and a little salt. This mixture is easily absorbed by the body and has saved many lives.

reduces the amount of water excreted passed out of your body) by the kidneys. At the same time you start to feel thirsty.

Blood filters

On average all the blood in your body passes through your kidneys once every five minutes. Blood enters the kidneys through the renal arteries and leaves through the renal veins. Inside each kidney the renal artery divides again and again into about 1 million tiny capillaries. Each capillary feeds blood into a miniature filter unit called a 'nephron'. There are about 1 million nephrons in each kidney. They filter the blood, producing a watery fluid that leaves the kidneys as urine.

Inside each nephron the capillary winds up into a knot called a 'glomerulus'. This knot restricts the blood flow and raises the pressure inside the capillary, forcing water and other substances to strain through the capillary walls into a tube called a 'tubule'. Many of the substances that pass through the walls, including glucose and salts, are still required by your body. These are absorbed back into the capillary, along with most of the water. The tubules from all the nephrons link up, finally merging into a larger tube called the ureter. The ureter carries the urine produced in your kidneys down to your bladder.

What is urine?

The major component of urine, apart from water, is a chemical called 'urea'. This is a waste product produced by the break-down of proteins in the body, for example in the liver. Your urine is darker when you have been drinking less because it is more concentrated. The yellow colour is produced by bile, a dark brownish-green liquid that also gives faeces their colour. Apart from urea and bile, urine contains small quantities of many other substances that your body needs to get rid of. These are usually invisible, but they can sometimes colour your urine or make it smell. For example, if you eat beetroot the chemical that makes the food purple will colour your urine red.

If your kidneys fail to work properly, toxic wastes build up in the blood. Patients with kidney failure can be treated by a process called 'dialysis'. Their blood

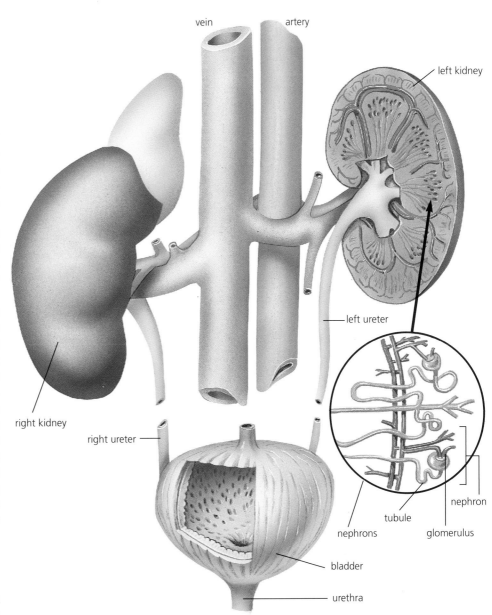

△ Your kidneys keep your blood clean. They are filters which sort and dispose of waste substances and excess water. Your bladder expands to store the liquid waste, then contracts to squeeze it down your urethra when you urinate.

supply is passed through a filtering machine which cleans the blood and adjusts its composition in a similar way to the kidneys. The cleaned blood then passes back into the patient's body. Dialysis must be performed every few days. It is expensive and time-consuming. Many people who have kidney failure receive kidney transplants. Sometimes the donor is a living relative of the patient, because the body of a healthy person can

function perfectly well with just one kidney.

We have seen how the water from a drink enters your body through the digestive system to join water that is already circulating in the blood. The circulation transports water to all the body's tissues, where it is needed to dissolve the substances such as glucose and minerals that support our life processes. We lose some of this water through our skin when we sweat, some when we breathe out, and some is filtered over and over again by the kidneys before eventually leaving the body in urine. Sensors constantly monitor all these processes, ensuring that the body's fluids are kept in balance. The story of a drink of water has more twists and turns than you might at first imagine!

REPRODUCTION

Reproduction is the driving force of life. A potential new life starts when a male's sperm cell unites with a female's egg cell inside her body. From this moment a unique new human being begins to grow.

All living things reproduce by passing on their genes (see page 13) from one generation to the next. Many simple organisms, for example bacteria, reproduce by dividing their bodies to make exact copies of themselves. The individual cells in our bodies reproduce in this way to produce growth, and to repair damage to our bodies throughout our lives.

The creation of a complete new human being, however, involves a much more complicated process. Human beings, like other complex organisms, reproduce sexually. In sexual reproduction a male and female mate and two special cells, one from the male and one from the female, join together. The moment when the two cells unite, or fuse, is called 'conception'. The new cell formed at conception contains genes (instructions) from both parents. The child which develops from it will have a mixture of the features of its parents.

After puberty (see page 72), when our bodies first become capable of reproducing, young men and women spend much of their time and energy finding a partner. If a couple decide to have children they will devote many years to looking after them, until the children become independent adults and perhaps have children of their own.

Sex cells

Both males and females have sex glands that are adapted to produce special sex cells. In males these glands are the testes, or testicles; in females they are the ovaries. The sperm cells and egg cells produced by these glands join together during mating. Females also have a special organ in which a baby develops – the uterus, or womb – and breasts for feeding milk to the baby after it is born.

Sex cells are different from all the other cells in our bodies because they contain only half the normal amount of DNA (see page 13). In a normal human cell there are 46 chromosomes (coiled strands of DNA and protein), which together contain all the genes that determine how our bodies are built. The 46 chromosomes in each of your cells can be arranged into 23 pairs. You inherited one chromosome of each pair from your mother and the other one from your father.

Dividing cells

Normally when a cell divides, each of the chromosomes in the cell replicates (copies itself). Each of the two cells formed by the division contains exact copies of all the 46 chromosomes that were in the original cell. This process is called 'mitosis'.

In your testes or ovaries however a special kind of cell division called 'meiosis' occurs. The special cells produced by meiosis contain only 23 chromosomes each, which is half a normal set. Also during meiosis the genes you inherited from your mother and father get shuffled or mixed up. This shuffling happens differently in every sex cell, so every sex cell contains a different combination of genes. This explains why you are not identical to your brother or sister.

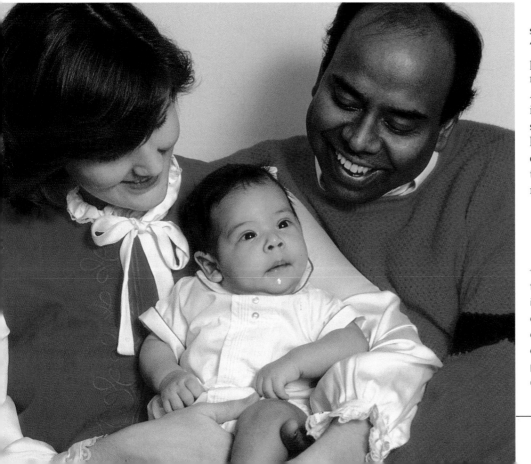

◁ Does this baby have her mother's eyes or her father's nose? Every new baby inherits a mixture of its parents' features in its genes. At the moment of conception 23 chromosomes in the father's sperm cell combine with 23 chromosomes in the mother's egg cell to produce a new combination of genes previously untried by Nature. The baby which develops is a unique new individual.

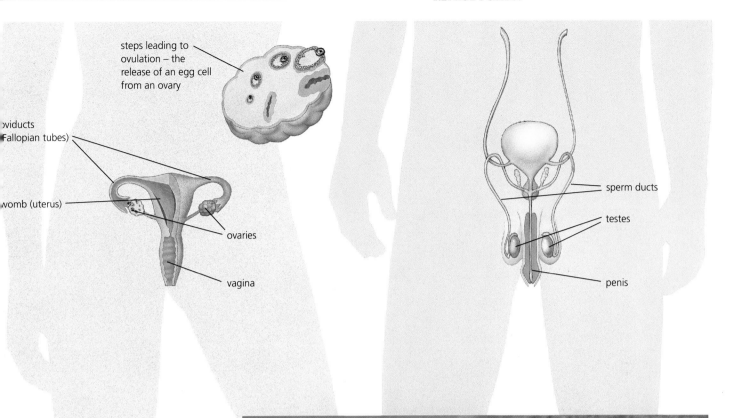

steps leading to ovulation – the release of an egg cell from an ovary

oviducts (Fallopian tubes)

womb (uterus)

ovaries

vagina

sperm ducts

testes

penis

△▷ Male and female sex organs are adapted to bring together the special sex cells which develop in the ovaries and testes. Sperm released by the male during sexual intercourse swim through the womb into the Fallopian tubes, where fertilization may take place. The fertilized egg makes its way to the womb, where the baby develops.

Egg meets sperm

If you are a girl you produced between 1 and 2 million egg cells while you were still in your mother's womb. After you reach puberty, at some time between the ages of 11 and 15, about once a month an egg cell is released from one of your ovaries. This is called 'ovulation'. The egg cell travels along a narrow tube called the 'oviduct', or Fallopian tube, towards your womb.

If you are a boy, at puberty your testes start to produce millions of sperm cells each day. Each sperm cell has a head and a tail which it uses to swim, like a tadpole.

During sexual intercourse a male inserts his penis into the female's vagina (see diagrams above). A fluid called 'semen' passes out of the penis and into the vagina. Semen contains millions of sperm cells which swim through the womb towards the oviducts. If an egg has been released from the ovaries and a single sperm manages to unite with it inside the

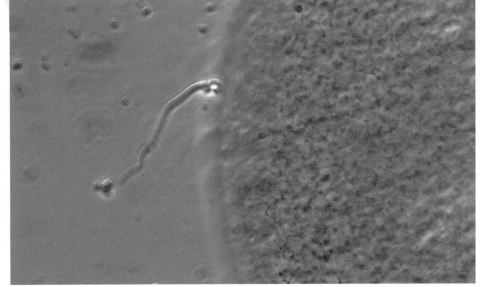

△ The moment of conception. A sperm burrows into the outer wall or membrane of an egg cell. Once the sperm is inside, the egg wall thickens to prevent any more sperms from entering.

oviduct, then fertilization takes place. If the egg cell is not fertilized, then after about 15 days it will be shed along with cells and blood from the lining of the womb. This is when a female has her monthly period.

When fertilization takes place the 23 chromosomes from each of the two sex cells combine to give a full set of 46. After about 12 hours the fertilized egg cell starts to divide, by mitosis, to form a ball of cells. The cells continue to divide about every 12 hours so that two cells become four, four become eight, eight become sixteen and so on. At this stage the cells are all identical. If for any reason the ball is split, any of the cells could continue to develop into a complete human being. Gradually the ball of cells makes it way into the uterus where, about five days after fertilization, it buries itself in the lining of the uterus wall. Now

Boy or girl?

Until very recently most parents had no idea if they were going to have a boy or a girl before their new baby was born. Now most parents in developed countries have seen ultrasound scans of their baby while it is still in the mother's womb. These scans are made to check the baby's health. They usually reveal the sex of the baby well in advance of the birth, although some people prefer not to be told.

The sex of a human being is determined from the moment of conception, when the sperm cell unites with the egg cell. The genetic information which told your body whether to develop into a boy or a girl was carried by your father's sperm.

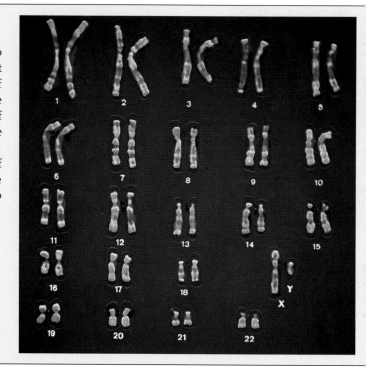

▷ The 23 pairs of human chromosomes. The two at the bottom right are the sex chromosomes. In this case they are an X and a Y, so the person is male. When pictures of a female's chromosomes are arranged in the same way the sex chromosomes are both X, and all 23 pairs of chromosomes match. The presence or absence of the Y chromosome makes a person male or female.

the dividing cells start to differentiate (develop in different ways). Part of the ball of cells becomes the placenta. This is a special organ which remains attached to the wall of the uterus during pregnancy. It collects oxygen and nutrients from the mother's blood and passes them to the growing baby. The outer layer of cells becomes a skin-like container, or sac, which protects the baby growing inside it.

Pregnancy

The female's first sign that she is pregnant is when her monthly periods stop. When her body detects the developing baby in her womb, it releases hormones to prevent further ovulation (egg release). These hormones trigger the changes to her body that will be needed to support the growing baby. From fertilization to birth the pregnancy will last for about nine months. During this time the baby will grow from a single cell too small to see without a microscope to a living being built from 100 million million cells. These cells have organized themselves into all the tissues and organs that make up a human body.

From implantation of the fertilized egg cell to eight weeks after conception, the developing baby is known as an 'embryo'. At eight weeks its nervous system and spinal cord have started to form, there are small pits (dents) where the eyes and ears will be, and a tiny heart.

After eight weeks the embryo is clearly a human being. It has fingers and toes and its external sex organs start to develop, showing if it is a boy or a girl. From this point on the developing baby is known as a 'fetus'.

Sixteen weeks after conception the fetus is floating in fluid inside its sac, held in place by the umbilical cord. Your umbilical cord was attached to your navel, or 'tummy button', when you were in your mother's womb. The other end of the cord is attached to the placenta. The fetus's

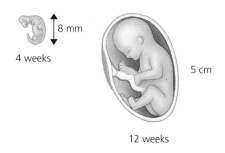

8 mm
4 weeks

5 cm
12 weeks

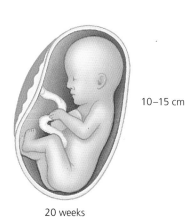

20 weeks

10–15 cm

▷ The fetus develops inside a fluid-filled sac cushioned from the bumps and jolts of the outside world. As the fetus matures it becomes more active and the mother can feel it turning and kicking. Some mothers say that their baby responds to sounds from outside the womb, soothing music for example.

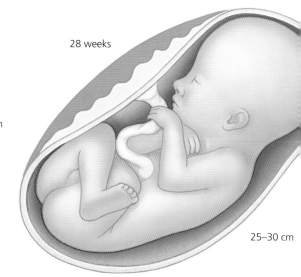

28 weeks

25–30 cm

blood circulates through this cord to the placenta, where it absorbs oxygen and nutrients from the mother's blood supply.

After 24 weeks (six months) hair has started to grow on the body of the fetus. It still weighs only 800 g but, with special care, from this point onwards it might survive if it was born prematurely (before the end of the nine months).

During the final three months of pregnancy the fetus puts on weight and gains strength. By the time it is ready to be born the baby will weigh about 3 kg.

Birth

Shortly before her baby is born the mother goes into labour. This consists of a series of strong, and painful, muscular contractions (tightenings) by the walls of the uterus as they try to push the baby out. Labour may last for several hours but, eventually, the baby is born. In most births the baby is delivered head first through the mother's vagina. If it is the wrong way around in the womb, or cannot pass through the vagina,

▷ Identical twins develop from a single fertilized egg cell which splits in two as it develops. They have identical genes.

a Caesarean section is needed. This is a surgical operation in which the baby is removed through a cut made in the wall of the mother's abdomen and uterus.

Soon after the baby is born, the placenta also emerges from the mother's vagina. It is still joined to the baby by the umbilical cord. The midwife clamps the cord with a special clip to prevent bleeding and cuts it to separate the baby from the placenta.

Twins and more

In the United Kingdom about 1 in 80 births produce twins. Three out of four sets of twins are non-identical. They may be boy and

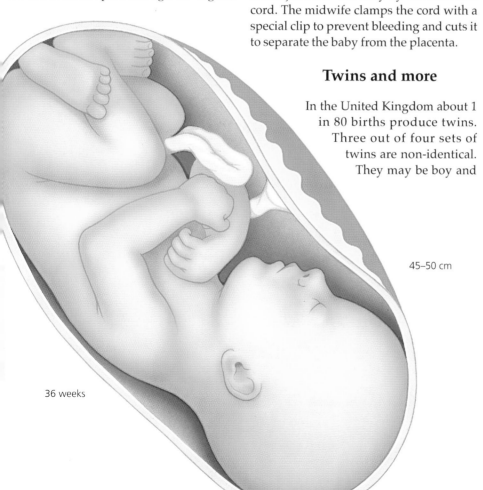

45–50 cm

36 weeks

girl twins, or both the same sex. Non-identical twins are conceived when two sperms fertilize two eggs which were released from the mother's ovaries at the same time. Apart from being the same age, non-identical twins are as similar, or as different, as any other pair of brothers and sisters.

About one in four sets of twins is identical. Identical twins are produced when an egg cell that has been fertilized normally splits in two as it starts to develop. Because identical twins are built from the same genetic plan, any differences between them must be produced by their surroundings and by their experiences of life as they grow up. In the rare cases where identical twins have grown up apart from each other, they have still developed very similar personalities and interests.

Very rarely a fertilized egg splits only partly as it develops. This results in Siamese twins. The bodies of Siamese twins are joined in at least one place.

Multiple births of three or more babies at once used to be very rare indeed. They became more common when fertility drugs were first used. These drugs are used to treat women who want to have a baby but who are not ovulating (releasing eggs) normally. They can make the ovaries release several eggs at once, though with modern fertility treatments this is now much less common. As a result of fertility treatment, several sets of sextuplets (six babies in one pregnancy) have been born. Imagine having six new brothers and sisters all in one go!

THE CYCLE OF LIFE

The life of every human being follows a common pattern from birth to death: infancy, childhood, the teenage years, maturity and, finally, old age. We will all experience each of these life stages, unless we die early as a result of an accident or illness.

Since the day you were born you have developed from a helpless baby, unable to look after yourself, into an independent human being. In that time your body has grown to several times its original length and to more then ten times its original weight. You have learned to feed and dress yourself, to talk, to read and write, and to find your way around a complicated world. You probably still live with your family, but in a few years' time you may leave home to attend college or to start a job. Eventually you may set up a home with a partner and have children of your own, continuing the cycle of human life.

Throughout our lives our bodies and behaviour change. The most rapid changes take place during infancy, childhood and adolescence – from the moment we are born to the time when we reach adult height and weight at the end of our teenage years. Then, for 40 or more years as mature adults, our bodies change more slowly. Finally, as we enter old age, our bodies start to change more rapidly again until, inevitably, they stop working altogether and we die.

The first year

Infancy lasts for the first year or so of life, during which a newborn baby gradually becomes aware of her surroundings. She starts to recognize her mother and, later on, other people. She learns to sit up, to crawl and perhaps to walk.

At birth the new baby has several reflex responses (see page 53). She sucks when something touches her mouth, blinks at bright lights and grips anything placed on her hand. She gazes at faces and follows movements with her eyes. All these new experiences are registered by the baby's developing brain. By the time she is 4 weeks old she recognizes her mother's face and smiles when she sees it. When she is 12 weeks old she turns to look towards the source of a sound. By 20 weeks she reaches for objects and studies whatever she is holding in her hand.

As the baby grows she becomes more able to control the movements of her body. By 6 months, she can lift her head to look around when she is lying down, and she can feed herself with a biscuit. She now starts to respond to her own name. By 10 months she will start to crawl, and by 13 months she may be taking her first steps. At the same time she will start to imitate sounds and understand some of the words she hears.

During this first year the baby's body grows more rapidly than at any other time in her life outside her mother's womb. By her first birthday she will be about 50 per cent taller than when she was born. If she continued to grow at this rate she would be over 2 m tall by the age of 5! Her rate of growth now slows down, and for the rest of her childhood her height will increase by about 5 to 8 cm each year and her weight by 2 to 3 kg.

◁ Before his first birthday this baby will change more rapidly than during any other period of his life. It is an exciting time!

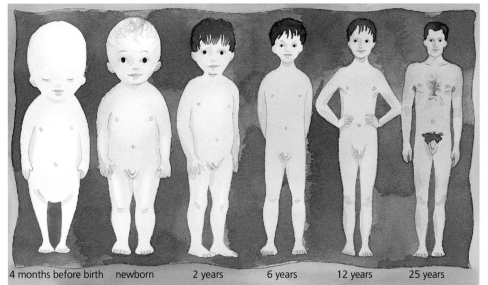

4 months before birth   newborn   2 years   6 years   12 years   25 years

◁ A baby has a large head in comparison to the rest of its body. Your proportions change as you grow. Between birth and adulthood your arms and legs grow most, and your head grows least.

Human beings have a longer period of childhood than any other animal. Rabbits are capable of breeding just 6 months after they are born, and horses mature when they are 2 years old. Only animals with large brains and complex social lives have a childhood of similar length to our own. Gorillas, for example, mature between the ages of 11 and 12 years. Our extended childhood has probably evolved to give us the time we need, under the care of our parents, to learn all the complex informa-

tion and skills that the human lifestyle requires. Human beings are born without the skills that distinguish us from other animals, for example language and the ability to use tools. Childhood is the time when we learn these special skills.

In early childhood we are still learning to control our bodies. Between 18 months and 3 years children stop wearing nappies and learn to use a potty. With the help and prompting of their parents, children gradually become aware of the sensations that signal that they want to go to the toilet, and they learn to wait for the appropriate moment. A three-year-old child can dress herself, stand on one leg for a short time and catch a ball.

Our first words are usually spoken before our second birthday and from then on we learn new words every day. At nurseries and playgroups we can socialize with other children of the same age, learning to wait for our turn and to share with others. The next big step is starting school.

◁ Starting to walk and talk – in early childhood there is something new to learn every day.

Childhood days

What can you remember from your childhood? For many people their childhood – the period between the first year of infancy and the teenage years of adolescence – is a magical time filled with play, discoveries, summer holidays, first school days, friends and toys. These years are spent growing and learning before our bodies start to mature.

▷ Learning to read and write during our first school years is the start of an education that may last for another sixteen years or more.

From child to adult

Adolescence is the period of life during which we take the step from childhood to adulthood. There are no definite dates on which adolescence starts and ends, but it is usually associated with the early teenage years.

During adolescence your body passes through the life stage of puberty. Your body matures sexually and you become capable of having children. If you are a girl, the changes in your body may start at any time from the age of 10 onwards. In boys they usually take place a year or two later.

At the start of puberty your body has a growth spurt. Because this spurt often happens earlier in girls than in boys, many of the girls in a class of 11- to 12-year-olds may be taller than many of the boys. By the ages of 14 to 15 the boys will, on average, have caught up with the girls, and from then on they will mostly be taller. You usually reach full adult height between the ages of 16 and 20.

The physical changes that take place during puberty are stimulated by the release of male and female sex hormones (see page 59). As well as making the sex organs themselves mature, these hormones make us develop secondary sexual characteristics, which change the appearance of our bodies. Boys and girls grow more body hair under their armpits and in the pubic region, and boys start to grow hairs on their chests and chins as well. The sebaceous glands, which produce oil (see page 32), become more active, particularly in boys, giving many teenagers skin problems such as spots and acne. The male sex hormone testosterone stimulates bone and muscle growth so that boys' bodies become heavier and stronger. As their larynx and vocal cords grow, boys voices break, becoming deeper and more powerful.

The sex hormones in girls start the regular monthly release of an egg from their ovaries, resulting in periods. Female hormones also stimulate the development of breasts in girls and build-up of extra fat under the skin. This makes girls' bodies more rounded at this stage.

These physical changes in teenagers' bodies are accompanied by emotional changes. During adolescence you become

△ During adolescence many young people first become aware of political issues and start to form their own views of right and wrong. Adolescence is frequently a time of conflict with parents and other authority figures such as teachers. These conflicts are almost inevitable as we make the transition from being children, totally dependent on adults for all our needs, to being independent adults responsible for our own actions.

much more conscious of your appearance to others than you were during childhood. You start being attracted to members of the opposite sex and begin to experiment with dating and relationships. Some people discover that they are attracted to people of the same sex.

Reaching maturity

By the time you are a young adult, in your twenties and early thirties, your body will be physically mature. This is the time of life when your body will be strongest and most agile. On the other hand if, during this period, you eat too many of the wrong foods and do not take enough exercise, you can lay the foundations for health problems that may shorten your life later.

Adult bodies change only relatively slowly. Beyond the mid-thirties the strength and speed of muscle tissue gradually decline. This is the age at which many professional sportspeople retire. Yet recreational sport can be enjoyed at any age. Many people have completed marathon races in their seventies and eighties. We may tend to put on weight as our level of activity declines, but active involvement in sports helps to avoid this and helps to keep our muscles and bones strong.

Men continue to be capable of fathering children throughout their lives. There are several recorded cases of men in their nineties fathering a child. Women, however, stop ovulating at about the age of 50. Over an interval of 1 to 2 years their monthly periods gradually stop. This stage in a woman's life is known as the 'menopause'.

After the menopause, a woman's body changes. From now on she is as likely as a man to suffer from heart problems. Her bones may become lighter and weaker. This bone loss could make her body stoop and her bones break easily in old age. However, if she continues to take exercise this will help to keep her bones strong. Some women take additional hormones after the menopause to help their skeletons retain their strength. This is called hormone replacement therapy or 'HRT'.

From the age of about 60 onwards, as their bodies begin to age people start to retire from full-time work. Most people retire by the time they are 65, though some professions including judges, politicians, actors and musicians often carry on working into their seventies.

Old age

Nothing lasts for ever and, like everything else, our bodies eventually wear out. In past centuries most people died as a result of accidents, violence or disease long before they reached what we now think of as old age. For example, in the United Kingdom at the start of the 20th century average life expectancy was about 50 years. Of course some people did live to a ripe old age – throughout recorded history there have been people who have lived for 90 or even 100 years. With modern medicine and improved living conditions, many more of us can now expect to live for 70 years or more. In developed countries the average life expectancy is now between 70 and 80 years. On average, women live about five years longer than men. This could be because of the protection that the female sex hormones give against heart disease and other illnesses before the menopause.

As our bodies age, our joints become stiffer and our bones more brittle. Our muscles lose their bulk, our height starts to decrease and our hair turns grey. Our skin loses its elastic properties and becomes more wrinkled. Inside our bodies the walls of the arteries become less elastic and thicker. This change, combined with the heart's reduced muscle power, means that our circulation delivers less oxygen

△ Physical capacity declines with age but mental abilities need not. An old person has a lifetime's knowledge stored in her memories. The wisdom of the elderly is greatly respected.

▷ If you are going to be an Olympic gold medallist you are most likely to achieve success in your twenties. This is when your muscle bulk, strength and speed will be at their peak.

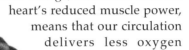

around our bodies. For this reason, older people are more likely to suffer from breathlessness, for example when climbing stairs. The most common cause of death in old people is heart failure.

Is there a limit to human life? Will people in the future live for 150 years or even longer? In fiction, the search for an 'elixir of life' that will make its discoverer live for ever is a common theme. Doctors are still not completely sure what makes our bodies grow old. It may be that a steady accumulation of accidents and damage from infections eventually wears our bodies out. If this is the case, then as we take greater care of ourselves we should expect to live longer and longer. But the limits to human life do not seem to be increasing in this way. Despite modern medicine it is still rare for someone to live much beyond the age of 100. Many doctors now believe that the ageing process is controlled by our genes. A limited number of cell divisions may be programmed into our genes when we are born, in which case there is little hope that we can extend our lives much beyond their present span.

HEALTH AND DISEASE

Your body has a variety of defence and repair systems that fight off everyday infections and put right minor damage. Modern medicine has developed effective treatments and cures for many serious illnesses that, until recently, were often fatal.

It is only really during the 20th century that we have understood the causes of most kinds of ill health. Before then medicine was often a mixture of myth and magic. Many 'doctors' had no scientific understanding of the human body, and their treatments such as taking blood from patients with blood-sucking leeches, or placing them in scalding hot baths, often did more harm than good. Modern medicine, based on scientific ideas, has shown how diseases are caused and, in many cases, how they can be cured. For example, we have discovered that infectious diseases – the ones you 'catch' – are carried by microbes (organisms that are too small to see without a microscope) that invade the body. The spread of these invading microbes, or bugs, can be reduced by simple precautions such as washing hands after going to the toilet, drinking clean water and handling food carefully. In the past, when the importance of such simple hygiene was not understood, serious infectious diseases such as cholera and typhoid were much more common.

Defence and repair

The human body can defend and repair itself. Many of the body's cells are replaced by cell division (see page 66) as they are damaged or grow old. Your skin, for example, heals itself when it is cut. But the ability of your body's tissues to regrow is limited. You cannot grow a new finger or arm if you lose one in an accident. Your nerve cells do not regrow at all, and if part of the brain is damaged, it is permanently damaged, although another part of the brain may take over some of its functions. The ability of tissues to repair themselves also decreases with age. Older people may take much longer to recover from illnesses and accidents than the young.

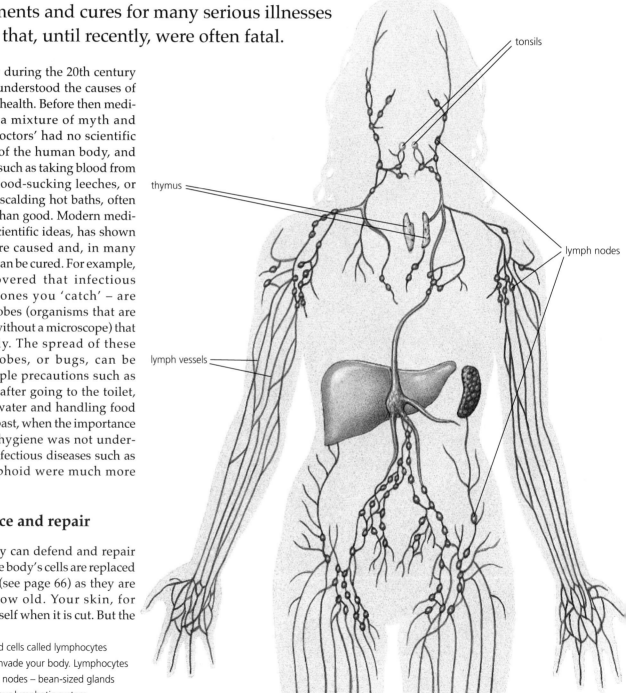

▷ Special white blood cells called lymphocytes attack germs which invade your body. Lymphocytes concentrate in lymph nodes – bean-sized glands which form part of your lymphatic system.

The body's first defence against invasion from foreign organisms is the skin. The skin is a barrier that microbes find difficult to penetrate. But they can get in through scratches and cuts and cause infections. They also enter our bodies in the air we breathe and the food we eat. Fine hairs and a slimy substance called 'mucus' in our air passages help to filter foreign bodies from air. Digestive juices in our stomach kill off many bugs in food. But when microbes succeed in crossing these defences, a protective reaction called the 'immune response' comes into action. This consists of your body's immune system recognizing and destroying invaders, giving you immunity (protection) against disease.

Fighting back

Your immune system is operated by white blood cells that are made in the marrow in your bones. Your circulation (see pages 42–47) transports these cells to all the body tissues. These defensive cells leave the blood vessels, along with nutrients and other chemicals, to form part of a clear liquid called 'lymph'. Lymph fills the spaces between the cells in your body. The level of lymph does not build up in the body because it drains back into the blood through a network of lymph vessels (tiny tubes similar to blood vessels).

When your body is invaded, lymphocytes known as 'B cells' produce special chemicals called 'antibodies'. These antibodies stick to the surface of the invading organisms, marking them so that they can be detected and gobbled up by phagocytes, another kind of white blood cell. Other lymphocytes called 'T cells' (because they mature in the thymus – see page 59) also take part in the defensive action. Helper T cells stimulate the B cells to produce antibodies. They also stimulate the killer T cells that attack invaders directly, often making them burst open. Suppressor T cells, another kind of T cell, stop the body's immune response from getting out of control.

Swollen tonsils

When you have an infection and your immune system is working hard, your lymph nodes may become swollen. You can feel them swelling in your armpits, in your neck or in your groin. Your two tonsils on either side of the entrance to your throat become swollen when you have a throat infection. Sometimes this makes it difficult for you to breathe and swallow. If you suffer from throat infections regularly, a doctor may decide to remove your tonsils. This was once a very common operation and your parents may well have had their tonsils taken out. Doctors are now more cautious about removing tonsils because they have a better understanding of their role in the body's defences.

ACCIDENTS AND FIRST AID

A knowledge of emergency first aid could help you to save someone's life in a serious accident. You can learn first-aid skills by attending a first-aid course given by qualified first aiders or paramedics.

At a first-aid course you will learn what to do if you are the first to arrive at the scene of an accident: how to spot the sources of continuing danger such as leaking gas or live electric wires; how to call for emergency help; how to deal with a casualty who has stopped breathing by clearing the airways and giving resuscitation (reviving them from unconsciousness); how to control bleeding by applying pressure; how to put an unconscious victim in the recovery position until an ambulance arrives; how to help someone who is choking or having a fit.

Learning these skills from an expert will give you the confidence to deal with emergencies both at home and when you are out and about with your friends.

▷ A student at a first-aid class is taught how to give the 'kiss of life'. If you want to become a lifeguard or water sports instructor then you must learn how to resuscitate a victim of a drowning accident.

Body invasion

Between 1347 and 1351 a disease called the bubonic plague – 'the Black Death' – swept through Europe, killing one quarter of the population. This dreadful disease starts with a high fever and produces terrible swellings of the lymph glands as the body tries to fight the infection. If the plague is not treated most victims will die after about four days. Outbreaks of plague have occurred across the world throughout history. The last great plague epidemic started in 1894 in China. There are still occasional isolated outbreaks today.

In medieval times the cause of plague was not understood. Most people thought that diseases were produced by evil spirits or the anger of God. Today we know that plague is caused by a kind of bacterium that enters the blood when a human is bitten by a flea from an infected rat. The insanitary conditions in medieval cities were ideal breeding grounds for the plague-carrying rats and their fleas.

Plague is an example of an infectious disease caused by foreign organisms invading the body and then multiplying inside it. Many different organisms, including viruses, bacteria, protists, fungi and animals, can infect the human body. These organisms are called 'parasites' – they survive by living inside our bodies and harming us at the same time.

Viruses and bacteria

When you have a cold your body has been invaded by a virus. Viruses are the smallest disease-causing agents. A single virus cannot reproduce itself when it is outside a living organism. This means that the virus itself is not really a living thing. The virus reproduces by injecting its genetic material (DNA or RNA) into a living cell, and then using the cell to make copies of itself. Eventually the cell bursts open, releasing many new copies of the virus to infect more cells.

Infections and diseases caused by viruses include the common cold, influenza, cold sores, warts, measles, mumps, rabies and AIDS. Some viral infections, such as the common cold and warts on the hands, are relatively harmless, and the body's defences eventually get rid of them. Others, such as rabies and the HIV virus that causes AIDS, can kill a person. The rabies virus is passed on when someone is bitten by an infected animal.

When you have a sore throat you have a bacterial infection. Bacteria are single-celled living things. Many bacteria are perfectly harmless (see page 9), but others cause disease. Typhoid, tetanus, food poisoning, boils, tonsillitis, pneumonia, tooth decay and meningitis are all caused by bacterial infections. Bacteria can be killed with heat and by antiseptics. Careful attention to hygiene helps to reduce the spread of bacterial infection.

Fungus, protists and animals

Have you ever had athlete's foot? This is an infection of the skin caused by a fungus. It results in itchiness between the toes and flaking skin. It is easily passed on from one person to another in changing rooms of swimming pools where the fungus thrives on the wet floors. Fortunately athlete's foot can be treated with a fungicidal (fungus-destroying) powder. Similar skin infections on other parts of the body are known as 'ringworm', though they are caused by a fungus, not by a worm.

Organisms called protists (see page 9) infect the body too. Water contaminated with protists called amoeba is a common cause of dysentery. One of the most widespread serious human diseases, malaria (see opposite), is caused by a protist which is spread by mosquitoes.

Humans can also be infected by animals such as mites, ticks, insects and worms. Fleas and lice are parasitic insects. More than 3 million children a year in the United Kingdom get head lice at school. These lay tiny eggs called nits in your hair. You can easily get rid of them with a special shampoo that contains a mild insecticide.

Worms and flukes enter our bodies from food or water contaminated with their eggs. Tapeworms and roundworms can live in human intestines. Their eggs enter our bodies through our mouths if our fingers or food are contaminated with faeces. Microscopic worms can be introduced into our eyes if we rub them after touching ground contaminated with dog faeces. These worms can cause blindness.

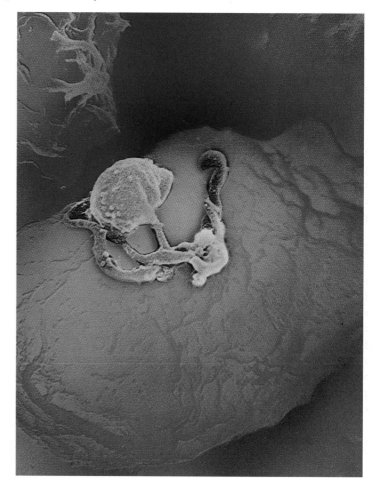

▷ The HIV virus that causes AIDS on the surface of a human helper T cell. The virus invades cells in the immune system, making someone who is infected much more vulnerable to other diseases.

THE MALARIA CYCLE

1 A female mosquito bites a human to suck a meal of blood before laying her eggs.

2 If the mosquito is carrying malaria, the malaria parasites are injected into the human's bloodstream. These parasites move to the person's liver, where they multiply in the red blood cells.

3 The red cells burst open, releasing spores (tiny reproductive cells) back into the victim's blood.

4 If another mosquito bites the victim it will pick up the spores which reproduce inside the mosquito's body, ready to infect the next human whom the mosquito bites.

▷ Malaria is a widespread disease in tropical countries. It probably accounts for at least one million deaths a year.

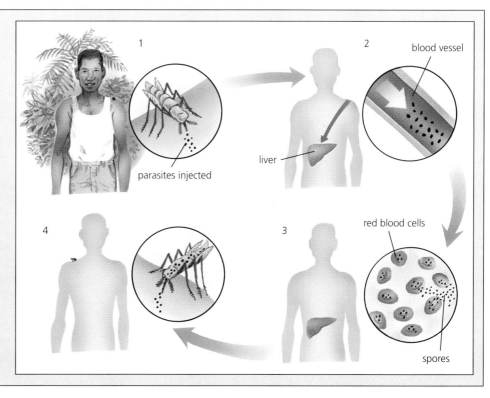

Fighting infection

The link between infection and disease was first discovered by a great French scientist, Louis Pasteur, in the 1850s. He was asked by the French government to discover what caused some wines to turn sour. Pasteur's experiments showed that wine sours when it is contaminated by a fungus whose tiny spores are present in the air. Wine can be sterilized by heating it to 55° C to kill any fungus present. This process of heating wine, or milk, to kill microbes is called 'pasteurization', after its discoverer. Most of the milk we drink is pasteurized to make it safer.

In England at the same time the surgeon Joseph Lister read about Pasteur's discoveries. In those days hospitals were not the sparkling clean, antiseptic places we are familiar with today. Surgeons often came straight from examining a dead patient and, without washing their hands, examined a living patient. Not surprisingly many infections were passed on. People were more likely to die from infections they picked up in hospital than from the injuries or illnesses that made them go there!

Lister realized that infections might be caused by invisible microbes being carried from patient to patient on the hands of doctors and nurses. He introduced new hygienic practices, insisting that hospital staff wash with carbolic (disinfectant) soap between examinations, sterilize instruments between operations and spray the operating room with antiseptic. At first his ideas were doubted, but his methods were a great success and infection rates in hospitals soon started to drop. Lister's ideas are now universally accepted and the greatest care is taken to avoid the transmission of infections within hospitals.

Sanitation

Pasteur's discovery of the role of microbes, or micro-organisms, in causing disease has had a great effect on our lives. Perhaps the greatest improvements in general health standards have taken place because we now realize that clean water, fresh air and

▽ Microbes causing infectious diseases thrive where there is no clean water or proper sewage system.

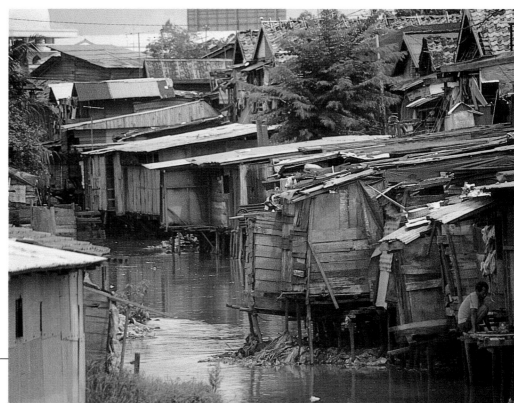

good living conditions reduce the risk of disease. In western countries water supplies are treated to destroy micro-organisms, and sewage systems keep human waste separate from the water we use for drinking, cooking and washing. Outbreaks of typhoid and cholera are rare when modern hygiene practices are followed. By keeping our environment clean and tidy we can keep down the numbers of pests, such as rats and flies, which transmit diseases to humans.

The milkmaid's story

The first successful vaccination to provide immunity against a disease was given in 1796 by an English country doctor called Edward Jenner. At that time a disease called smallpox used to kill tens of thousands of people every year. Jenner had survived smallpox as a child, and he knew that people who had caught smallpox once never caught it again.

As a young doctor Jenner treated a milkmaid who had caught cowpox from the cows she milked. Cowpox is a similar disease to smallpox, but much less serious. The milkmaid told Jenner that people who had suffered from cowpox never caught smallpox. This idea stuck in his mind, and later in his career he tried an experiment. He took liquid from the cowpox spots on a milkmaid's hands and scratched it into the arm of a healthy eight-year-old boy called James Phipps. Sure enough, James developed cowpox. Six weeks later Jenner scratched some material from the body of a smallpox victim into James. This was a very risky experiment because James might have caught smallpox and died. In fact the experiment was a success. James did not catch smallpox because he had already had cowpox, and so he was immune to the smallpox infection.

Vaccination

We now understand how vaccines work. A vaccine contains dead or harmless versions of the disease-causing organism. These do not produce the illness (though they may produce some of its minor symptoms), but they stimulate the body's immune system to produce the same defensive cells and antibodies that the active organism does. These cells and antibodies stay in the bloodstream, ready to defend the body if a real infection takes place. Following some vaccinations you may need a booster dose after several years to make sure that your army of defenders remains at full strength.

Vaccinations have greatly reduced the number of deaths from infectious diseases. In 1977 the last ever case of smallpox was recorded. A worldwide programme of vaccination has completely eliminated this disease.

Eighty years after Jenner's first vaccinations a scientist called Robert Koch developed an artificial vaccine against anthrax, a disease transmitted to humans by sheep and other animals. Following Pasteur's discovery of the role of microbes in causing infections, Koch cultured (bred) the bacteria which cause anthrax in a flask. He then heated the culture to kill the

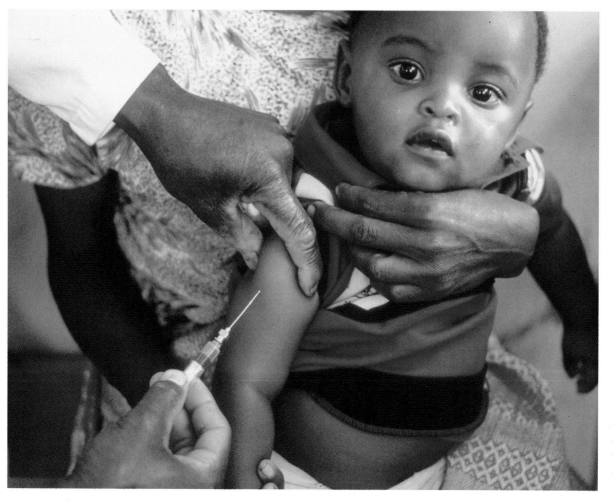

◁ Today most children are offered a series of vaccinations, usually by injection, which prevent them from developing potentially serious infectious diseases such as diphtheria, whooping cough, polio, tetanus, tuberculosis, measles and rubella (German measles).

bacteria. The sterilized culture was an effective vaccine against the disease.

The anthrax vaccine was the first in a long line of vaccines and other medical drugs produced as a result of scientific understanding.

Drugs versus disease

Through history people have taken potions and pills, often manufactured from herbs and other natural products, to treat their illnesses. The success of these treatments was often the result of a process of trial and error. Some were effective: for example, the Romans knew that a potion made from willow tree bark could help relieve pain. We now know that the willow contains a chemical similar to the modern drug aspirin. Foxgloves contain a substance called digitalis which can be used to stimulate the heart. But often these herbal remedies were used without scientific understanding and caused more harm than good, sometimes poisoning the patient.

Antibiotics

Artificial vaccines were one of the first great breakthroughs in the search for scientific medical treatments. Another breakthrough was the discovery of antibiotics – substances that kill bacteria.

The first successful antibiotic was penicillin. In 1928 a Scottish scientist called Alexander Fleming noticed that some dishes on which he was culturing bacteria had started to go mouldy. Fungus spores had landed on them and started to grow, just as mould grows on old bread. Fleming observed that all the bacteria around the mould growths had died. He realized that something in the mould had an antibacterial action. Fleming made a purified broth from the mould, which he called penicillin. He thought it might be a good antiseptic that could be painted on to wounds. It was not until 10 years later that two other scientists, Howard Florey and Ernest Chain, showed that penicillin could be swallowed to treat bacterial infections. Before antibiotics were available, infections in wounds often spread, leading to blood poisoning and gangrene (when the infected tissue dies). Now these infections can be treated with antibiotic drugs.

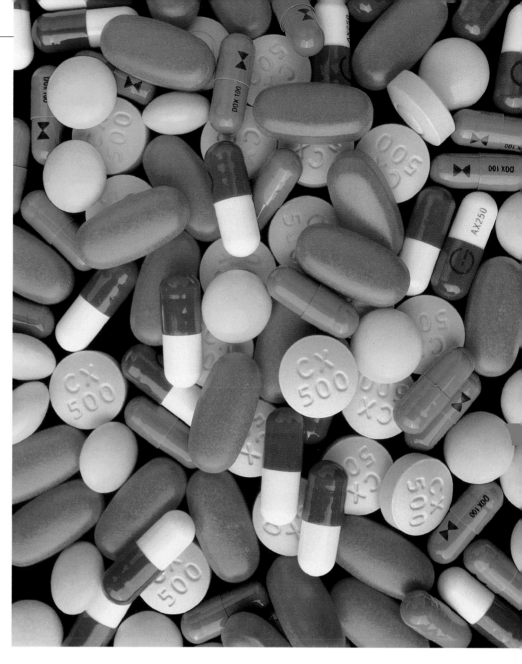

Wonder drugs

The number of drugs available to us today is enormous. Every year, thousands of new chemicals are manufactured and tested for possible beneficial effects on our bodies. Drugs such as aspirin and paracetamol, and more powerful substances such as codeine and morphine, are called analgesics (pain relievers). Many of these drugs work by affecting the way nerve signals are transmitted from the source of the pain or inside the brain.

Steroid drugs work like hormones to reduce inflammation (swelling) in damaged joints, to relieve the symptoms of asthma or to treat skin problems. Antihistamine drugs are used to treat allergies. You might take an antihistamine

△ Just some of the many different antibiotic drugs that are now available to fight bacterial infections. Your doctor may give you an antibiotic drug if, for example, you have a throat or chest infection.

tablet if you suffer from hay fever. It blocks the release of the chemicals in your immune system that produce the allergy symptoms. Anaesthetics can be used either locally in one part of your body, as at the dentist, or generally to put you to sleep so that you are not conscious of the pain of an operation.

All drugs are powerful substances that affect the body in several ways at once. Every drug has side-effects; for example, many people have an allergic reaction to

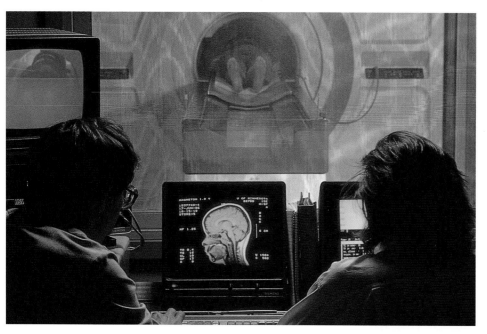

of surgery than from the more drastic surgical techniques used previously.

Perhaps the most dramatic operations performed today are organ transplants, in which a patient's kidneys, heart or lungs are replaced by organs from a donor. The greatest problem with the first transplant operations 30 years ago was that the patient's immune system identified the new organs as foreign invaders and attacked them in the same way as an infection. Now transplant patients are given special drugs to control their body's immune response. People can lead full and active lives with their transplanted organs.

Now doctors are experimenting with artificial organs. For many years, artificial hip joints have been used to replace worn-out hip bones. The first artificial hearts are now being tested.

penicillin and develop a rash when they take it. Even taking aspirin can produce internal bleeding in some people. Doctors must make a careful judgement that the good done by a drug outweighs any harm it might cause. Some modern anti-cancer drugs, for example, make patients feel extremely ill – they lose their hair and feel nauseous while they are taking the drugs. Yet if the drugs cure the cancer, then it is probably worthwhile suffering their side-effects.

△ Modern doctors have many scientific aids to help in their diagnosis of illness. Blood tests, X-rays and body scans give detailed pictures of the condition of the body's cells and systems.

Surgery

When neither treatment with medicines nor the body's own defences can solve a health problem, doctors may resort to surgery to rebuild damaged organs or remove infected tissues. Surgery places great strain on the body and is only performed when really necessary. Before this century surgery was crude and brutal. Patients received no anaesthetic and often died from shock or infection. These days, however, surgery is highly scientific and refined. Operating theatres are completely sterile. Instruments that use optical fibres (bundles of fine glass threads that transmit light) allow surgeons to see inside our bodies. The instruments are inserted into the patient through tiny keyhole incisions (cuts). Surgeons can even operate through the keyhole using tiny instruments attached to the ends of wires. Patients recover much more quickly from this kind

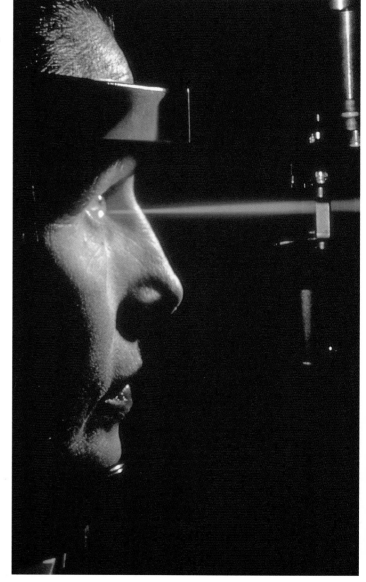

▷ Delicate eye surgery is performed with a laser beam. With this technique much more accurate incisions (cuts) are possible than with a hand-held scalpel.

The fight against cancer

About one in five people in developed countries dies from cancer. Cancers occur when the chromosomes in some body cells become damaged or altered and the cells start to divide unusually rapidly to produce a tumour (growth). Some tumours are not a cause for concern: they simply form a lump that stays in one place in the body. These tumours are described as 'benign'. A wart is a benign tumour on your skin, caused by infection by a virus. Other tumours are malignant, or cancerous. As a malignant tumour grows, cancerous cells break away from it and travel around the body in the blood supply or lymphatic system. These cells produce new tumours. The growth of the tumours damages organs and tissues, and weakens the body's immune system. The patient picks up infections and wastes away.

A cancer tumour may form in any part of the body, but it is most likely to start where the cells normally divide rapidly, for example in the skin, in the lining of the intestines, in the lungs or in the reproductive organs. Many factors are involved in producing the damage that makes cells become cancerous. Toxic (poisonous) chemicals can produce cancer; for example, the vast majority of people who develop lung cancer are smokers. Tar in the smoke they inhale damages cells in the lining of their lungs. Cancers may also be

triggered by some viruses, such as the hepatitis B virus. Radioactive fall-out from nuclear weapons and from accidents at nuclear power plants produces cancer. Many people exposed to the large amount of radiation released from the Chernobyl accident in the Ukraine in 1986 may develop leukaemia (cancer of the blood cells). Too much exposure to ultraviolet radiation from the sun can produce skin cancer, particularly if your skin is very fair. Diet may also be a factor in causing cancer.

Some cancers which were once always fatal can now be treated successfully. Leukaemia in children, for example, can be treated with powerful drugs that kill the cancer cells. This method of treatment is known as 'chemotherapy'. If tumours are spotted early they can be removed by surgery before they become malignant. Cancer cells can also be killed by radiation therapy, where special machines concentrate powerful doses of radiation on to the tumour.

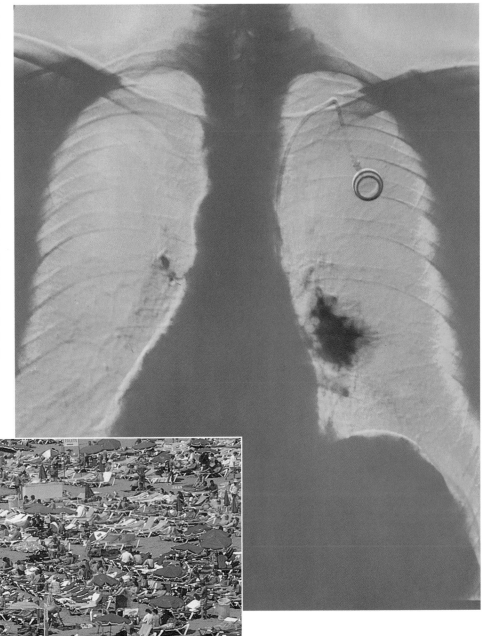

▷ The dark red mass in this patient's left lung is a cancerous tumour. Most lung cancers are caused by cigarette smoke.

◁ As the fashion for holidays in the sun and a 'healthy' tan has grown, so has the number of people developing skin cancer. People with fair skin must take particular care not to burn.

Hereditary disease

In some parts of Africa up to one in five people have inherited a gene that causes their red blood cells to be deformed. The cells are thin and are shaped like a crescent moon or sickle, hence the name of the disease – sickle cell anaemia. If a person has just one copy of this gene then only some of the blood cells are affected and the person is healthy. But if they inherit the gene from both parents, then all the blood cells will be affected and the person will suffer from sickle cell anaemia. Many people who have this disease die in childhood.

Sickle cell anaemia is a hereditary disease. Hereditary diseases are caused by faults in the genetic information carried by our chromosomes (see page 66). We have 23 pairs of chromosomes in each of our cells. If only one chromosome in a pair carries the faulty gene, then the normal gene on the other chromosome can do the required job. In this case the person is a 'carrier' of the disease but does not develop it themselves. The faulty genes are passed on from one generation of a family to the next.

Some other common inherited diseases are: cystic fibrosis, in which the sufferer's airways produce too much sticky mucus and become blocked; spina bifida, which produces deformations of the bones; muscular dystrophy, which causes the sufferer's muscles to waste away; and haemophilia (see below).

Genetic tests, which read portions of a person's genetic programme, or DNA (see page 13), can now warn people if they are at risk of passing on faulty genes to their children. Some genetic illnesses can be treated by replacing a substance that the faulty genes fail to manufacture correctly. For example, haemophilia is treated with regular injections of blood-clotting substances from donated blood. But there are still no cures for many other hereditary diseases.

Gene therapy

One possibility for the future treatment of inherited disease is the development of so-called 'gene therapy', in which non-faulty versions of the faulty gene are introduced into the patient's DNA. This could be achieved with a specially prepared virus that injects the DNA into the patient's cells. Gene therapy is in a very early stage of development, but it holds out great hope for curing previously untreatable diseases.

◁ Queen Victoria (seated at the front) was a carrier of the hereditary disease haemophilia. Only men develop the disease. Their blood does not clot and they can bleed to death from a simple cut or bruise. Victoria, who had nine children and 34 grandchildren, passed on the faulty gene to many of her descendants. One of her daughters and two of her granddaughters who inherited the gene appear in this picture.

Healthy choices

The progress in medicine in the past century has been staggering. Vaccinations and new standards of public hygiene have almost wiped out many terrible diseases from the developed world. New drugs treat infections, relieve pain and fight cancer. Damaged organs can be repaired or replaced by delicate surgery.

We must take care of our bodies to get the best from them. Many of the illnesses which affect people in wealthy countries are the result of their choice of lifestyle. Perhaps the best understood link between a choice we make and our future health is the effect of smoking. The latest studies show that at least one in two people who choose to smoke will die early as a result. A poor diet, stress and lack of exercise, combined with the use of alcohol, tobacco or other drugs, increase the risk of heart disease and other conditions. The abuse of substances such as solvents and illegal drugs can produce a temporary 'high' for the individual, but carries the risks of death from an overdose, poisoning, addiction and associated health problems.

Many people in developed countries are fortunate to have the opportunity to make these choices. The poor, and the populations of poorer countries, however, are not so lucky, and their health often suffers as a result.

▷ A healthy lifestyle includes a balanced diet, regular exercise, a comfortable place to live and stimulating work and recreation. By making the right choices we can increase our chances of enjoying long active lives.

▷ Living in a polluted environment has damaged many people's health. As industrial activity increases around the world we must ensure that factories and power stations do not release poisonous chemicals into our air and water.

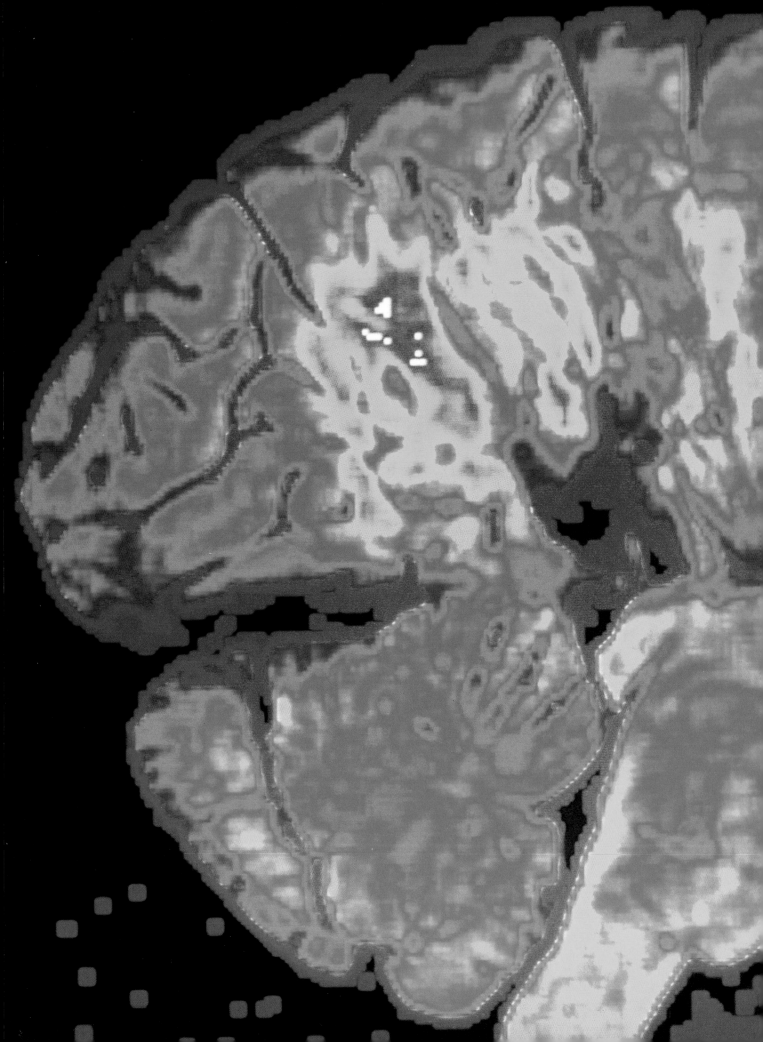

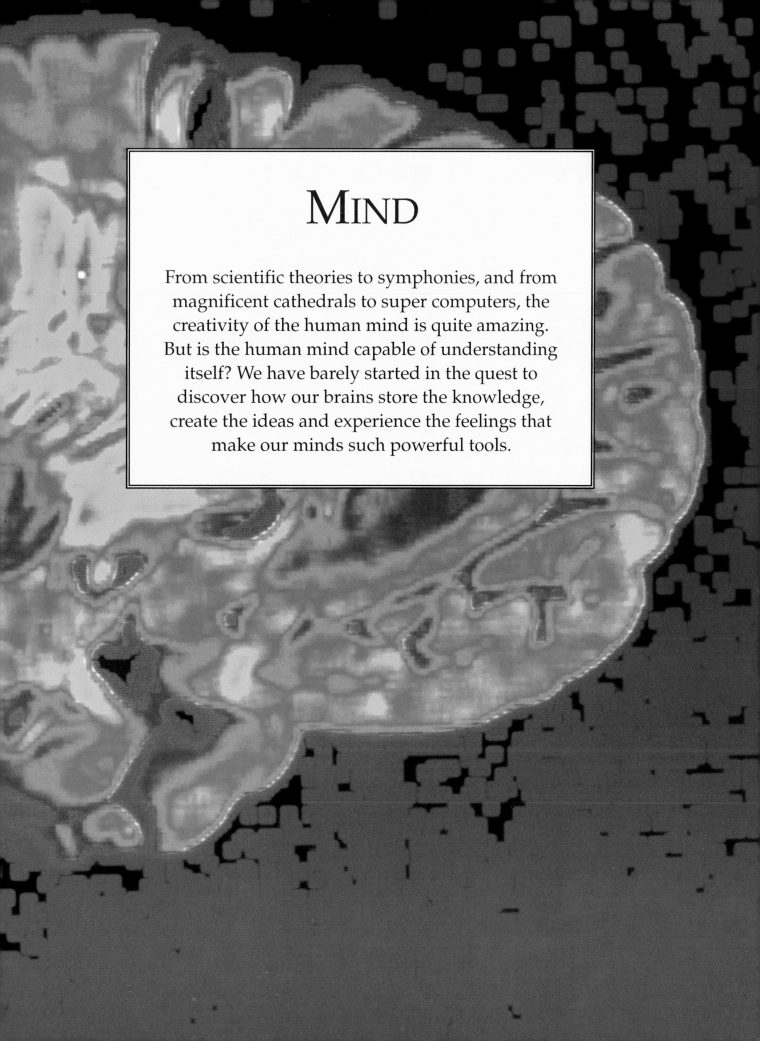

MIND

From scientific theories to symphonies, and from magnificent cathedrals to super computers, the creativity of the human mind is quite amazing. But is the human mind capable of understanding itself? We have barely started in the quest to discover how our brains store the knowledge, create the ideas and experience the feelings that make our minds such powerful tools.

THE BRAIN

When you look at a familiar face, how does your brain use the signals it receives from your eyes to create a picture of that face in your mind? How can you recognize the face and then associate it with a name? The answer is we do not know!

As far as we know the human brain is the most complex machine in the universe. We know about some of the things it can do, but exactly how does it create your sensations, memories, thoughts and feelings? Most mysterious of all, how does it know that it is doing all this? Understanding the brain is one of the great scientific challenges of the future.

The human brain is often compared to a computer. Like a computer, it is a multi-purpose machine that is able to perform an endless number of tasks: it controls your body, stores your memories, solves problems and produces language. But it might be better to compare your brain to a network of computers that work together to keep your body running smoothly. Different regions of your brain perform different tasks at the same time. However, unlike a computer, your brain is aware that it is doing all this information-processing work. It is conscious.

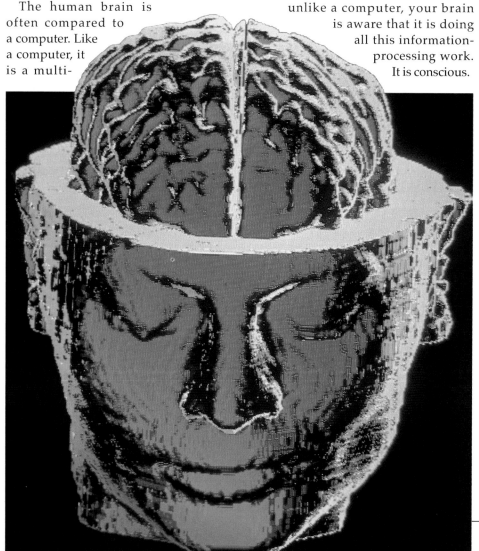

Building a brain

The science of the brain is in its very early stages. We have discovered some of the things that different parts of the brain do, and what can go wrong when they are damaged. We also understand the building blocks from which the brain is made – the neurons (nerve cells, see pages 52–53) – and how individual ones work. You have more than 100 billion neurons in your brain, which is about the same number as the number of stars in a galaxy.

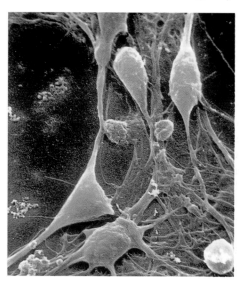

△ The brain's neurons branch and connect to form the most complex information-processing network on the planet.

Your brain needs energy to work properly. Your brain cells are fuelled by glucose and oxygen, which are delivered by your circulation (see pages 42–47). When you study hard you burn up more calories than when you let your mind go 'blank'. Studies have shown that pupils who have a good breakfast can concentrate in school for longer than those who skip breakfast.

Has anyone ever told you to 'use your grey matter'? They mean, think! The external surface, or cortex, of your brain, consists mainly of grey material. It is made up of neuron cells. Some internal parts of the brain are made of white matter. This consists mainly of the nerve fibres (axons) which extend from the neuron cells. These axons carry signals from one part of your brain to another.

◁ The surface of the cerebral hemispheres, where your thoughts and feelings are produced.

Signals

Each neuron links to hundreds, or even thousands, of other neurons in your brain. Some neurons are connected to neighbouring neurons, but others link to neurons in more distant parts of the brain. A neuron sends out signals that travel as a series of electrical pulses. They move along the axon to the end, where it branches. The tips of the branches make contact with other neurons. These contact points are called 'synapses'. The electrical pulses set off the release of chemicals called 'neurotransmitters', which carry the signals across the gaps between the neurons to the receiving neuron.

The rate at which a particular neuron sends out pulses depends on the rate at which it receives signals (impulses) from other connecting neurons. Some of these signals stimulate the neuron to pulse; others slow down its activity. Since a neuron might receive signals from hundreds of other neurons at once, the output of each neuron is like the answer to a complicated addition sum.

In most modern computers, information is in the form of streams of electrical pulses. Numbers, letters, sounds and pictures are stored by millions of tiny electronic switches. In a similar way, everything that happens in your brain – the impulses that stimulate your muscles, the signals from your senses that create pictures and sounds in your mind, even your thoughts and feelings – is produced by electrical and chemical pulses in your neurons. Yet the circuits in your brain are far more flexible than those in any computer. As your brain grows and develops, the connections between neurons change according to your experiences. Your brain is like a computer that is constantly rewiring itself!

Brain parts

Inside the skull, three distinct parts of the brain are visible: the brain stem, which connects the brain to the spinal cord and disappears into the base of the brain; the cerebellum at the base of the brain, behind the brain stem; and the cerebrum. The cerebrum consists of two distinct halves which are separated by a deep groove. These are the left and right cerebral

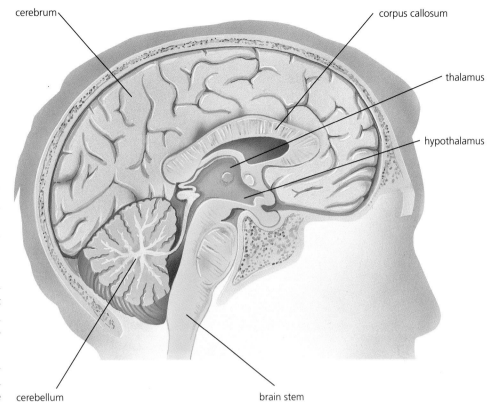

cerebrum — corpus callosum — thalamus — hypothalamus — cerebellum — brain stem

△ A cross-section through the centre of the brain, showing the separate structures from which it is built. Each part of the brain is specialized to perform one or more tasks.

hemispheres. They have a crumpled appearance like screwed-up sheets of paper. The hemispheres are made up of folded sheets of nerve tissue which, if unfolded and spread out, would be large enough to cover a small table. They surround lots of other brain parts which would be exposed if the brain were cut in two. These parts include the hippocampus (see page 88), the thalamus and the hypothalamus.

Mapping the brain

Accidents as well as illnesses that cause damage to the brain, such as strokes (a burst blood vessel in the brain) have allowed scientists to work out what certain areas of the brain do. For example, the French surgeon Pierre Broca examined the brain of an intelligent patient who, before his death, had been unable to speak and write but could still understand language. Broca discovered that an area about the size of a golf ball in the patient's left cerebral hemisphere was damaged. This area of the brain is now called 'Broca's area', and it seems to be the area where our ability to use language is located.

Observations during brain surgery have also helped scientists to map the brain.

Surprisingly it is possible for patients to remain conscious during a brain operation. This is because there are no pain-sensitive nerve endings inside the brain. If part of a patient's brain is stimulated with a mild electric shock during surgery, the patient can describe what he or she feels.

WHAT'S IN A NAME?

The names of some brain parts sound very impressive and technical. In fact they are just descriptions of the shape or position of the various parts in another language – either in Latin or Greek. Cerebellum, for example, simply means 'little brain'. Cortex means 'bark' or 'rind', so the cerebral cortex is the 'brain rind', or the brain's outer layer. Hippocampus means 'sea horse', which refers to its shape. Thalamus means 'inner room', and hypothalamus means 'below the inner room'.

WHAT DOES IT DO?

The largest part of your brain, which gives it its distinctive 'cauliflower' appearance, is the **cerebrum**. The outer layer of the cerebrum is the cerebral cortex, which seems to be responsible for your conscious sensations and decisions.

It is your cortex that makes you aware of the sights, sounds, smells, tastes and touches of the world around you. It thinks your thoughts, solves problems, make plans and understands language. Also, it is in the cortex that you would design a motor car, write a piece of poetry, appreciate music and have your dreams.

The **limbic system** surrounds the centre of your brain. It seems to be involved with basic behaviours such as feeding, fleeing danger or controlling aggression.

The **hippocampus**, which is part of the limbic system, has an important role in forming memories. If you damage the hippocampus you do not forget memories of old friends and events, but you cannot form new memories.

The **thalamus** sits on top of the brain stem at the centre of your brain. It is a kind of telephone exchange, routing nerve signals to different parts of the brain. In front of the thalamus, and beneath it, is the hypothalamus (see diagram on page 87). This links your nervous system to your endocrine system (which produces hormones) and plays an important role in controlling the temperature, water balance and other vital conditions of your body, such as feelings of hunger and your daily cycle of waking and sleeping.

The **cerebellum** controls automatic muscle actions, such as balancing your body as you walk or ride a bicycle. If the cerebellum is damaged in an accident then body movements become jerky and uncoordinated.

The **brain stem** is the information highway between your brain and the rest of your body. It is a massive bundle of nerve cells which carry signals to and from your brain.

Scientists can investigate the activity of different areas of the brain with body scanners. Using an instrument called a PET (Positron Emission Tomography) scanner, scientists are able to observe which areas of the brain are activated when someone talks, listens to music or solves a problem. The person swallows a drink containing glucose which has been made slightly radioactive. The active areas of the brain absorb this glucose to fuel their activity. The scanner uses the rays given out by the radioactive glucose to create an image that shows where the activity is located in the brain.

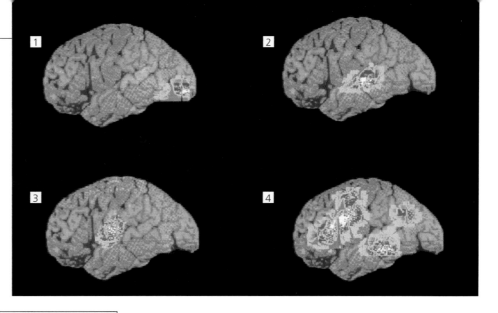

△ PET brain scans show which regions of the brain are active when someone is **1** looking at something, **2** listening, **3** speaking, **4** thinking.

PHINEAS GAGE

In 1848 an American railway worker called Phineas Gage had a terrible accident. An iron bar was being used to push a gunpowder charge into a hole. The explosive detonated unexpectedly and the iron bar was blasted from the hole and shot straight into Gage's head. It entered his skull below one eye socket and came out above his forehead.

Incredibly, Phineas Gage survived the accident and lived for another 12 years. But his personality changed after the accident. Before, he had been an easy-going and calm person, but afterwards he was irritable and bad-tempered. The iron bar had destroyed a large part of the front of his cerebral cortex. This unfortunate accident shows that the front of the cortex is concerned with forming our personality.

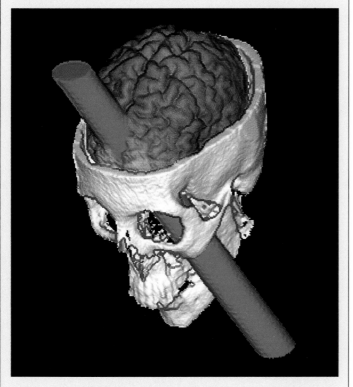

△ A computer model of Phineas Gage's head shortly after his accident.

▽ Some mental processes seem to be located in one side of the brain. Your left cerebral hemisphere can solve crossword puzzles or mathematical problems. Your right cerebral hemisphere is better at improvising music.

Left or right?

Brain-mapping studies show that some functions of the brain, for example voluntary (controlled) movements and the senses of sight, smell and touch, are located in both the left and the right sides of the brain. The nerves that produce sensations from, and send signals to, the left and right sides of the body cross in the brain stem. This means that your left brain hemisphere controls the right side of your body, and the right hemisphere controls the left side. When you stroke a cat with your right hand, for example, the signals that control your hand and the soft sensations of the fur come from, and travel back to, the left side of your brain. If someone suffers a stroke in the right side of the brain, the left side of the body will be affected. The person may have difficulty in controlling the movements of the left arm, for example.

Other brain functions take place on one side of the brain only. For example, Broca's area, which controls speech, is located in the left hemisphere. The parts of your brain that you use to work with numbers also seem to be on the left side. The right side of your brain does not understand language, but it seems to be good at understanding shapes, movements and working with patterns. People sometimes say that the left side of the brain is the verbal or analytical side, while the right is the non-verbal or creative side. When you have a discussion with someone, or try to work out a problem, you use the left side of your brain; when you sketch a picture or improvise at the piano you use the right side. About 90 per cent of people find it easier to use their right hand for tasks such as writing than their left hand. Of the 10 per cent who are left-handed, the majority still have speech located on the left side of their brain. In about 3 to 4 per cent of people the right hemisphere is dominant for speech as well as hand movements.

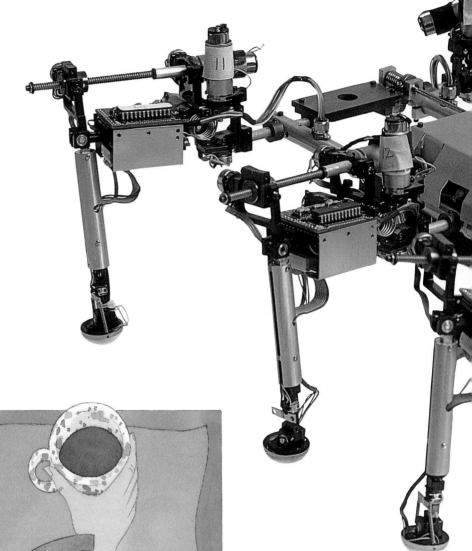

◁ A split-brain person can name the object in her right hand but not the one in her left. (The left side of the body is connected to the right, non-verbal side of the brain.) A screen blocks her view so that her right eye, connected to the left, verbal side of her brain, cannot see the object in her left hand.

Split brains

Normally the left and right sides of the brain are linked with a thick bundle of nerve fibres called the 'corpus callosum'. These fibres allow the two sides of the brain to communicate with each other. We are not aware that different processes are taking place in different parts of our brains. Epilepsy is an illness in which random brain pulses spread throughout the brain, causing the sufferer to have a fit. Patients with severe epilepsy are sometimes treated by having their corpus callosum cut. Patients who have had this

speech

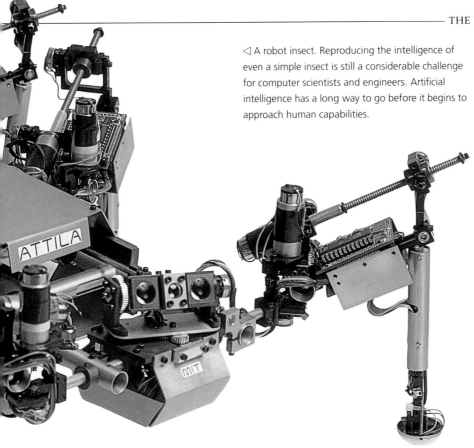

◁ A robot insect. Reproducing the intelligence of even a simple insect is still a considerable challenge for computer scientists and engineers. Artificial intelligence has a long way to go before it begins to approach human capabilities.

our brains, computers have inputs and outputs, sensors and memories, and they process information. But a computer has to be programmed by a human being. It has no sensations or feelings; it is not interested in the work it does, yet it does not become bored. Computers cannot feel pleasure or pain. They are better than us at some jobs, for example repeating the same task over and over again without making a mistake. But they are bad at doing some things that we find easy, for example recognizing a face.

Computers are becoming more and more powerful. Some are being designed with circuits that behave more like the neurons in our brains, but will they ever be able to do what our brains can do? Could an artificial brain ever be aware of its own existence? Could it feel happy or sad, enjoy music or warm sunshine? Do you think these same abilities can ever be created artificially in the laboratory?

▽ Are the circuits in a computer microchip really capable of behaving like a human mind? In the future we may discover if human intelligence can be reproduced electronically.

operation have a 'split brain'. It is almost as if they had two separate brains with different abilities.

How do brains really work?

Psychologists and neurologists are scientists who study the brain. Psychologists examine how our brains solve problems and how we respond to different situations and stimuli. Neurologists study the structure of the brain and the way in which our neurons operate. They understand how individual neurons work and in which areas of the brain our different mental processes are located. Yet they cannot explain how the electrical and chemical pulses of neurons create the sensations, memories, thoughts and feelings which psychologists investigate. We need information from both groups of scientists to understand the brain fully.

The use of computers to carry out jobs that are normally done by people is known as 'artificial intelligence'. Some scientists believe that the creation of artificial intelligence with computers will help us to understand how our brains operate. Like

PERCEPTION

You see a friend, hear a favourite song or brush against a soft animal. Your brain is constantly trying to make sense of daily experiences like these. Its job is to interpret signals that arrive all the time from your sense organs. Making sense of these signals is called 'perception'.

When you look at a person's face, each of your eyes forms an image of the face on its retina (see page 54). A camera forms an image on film in a similar way. But the camera film does not know that this is the image of a person; it does not recognize the person's face or remember their name. The film cannot be conscious of the image. Perception takes place in your brain, not in your eyes – after all, you 'see' things in your dreams when your eyes are closed! Perception occurs when your brain interprets signals from your senses or, in the case of dreams, images and other experiences recalled from your memory.

◁ Your brain perceives these cleverly arranged fruit and vegetables as a familiar object – a human being. Would a slightly less complex brain, a chimpanzee's for example, see the same thing?

Where is it?

When you look at a scene, your eyes detect the colour and the brightness of the light in the different parts of the image. But it is your brain that organizes the signals so that you are aware of the different objects before you. Camouflaged animals are sometimes difficult to spot; a flat fish, for example, merges into its surroundings or the seabed. Yet once you know the fish is there, you can use tiny clues to see its shape. Your brain organizes the various shapes and colours into patterns you recognize.

△ Can you see this camouflaged stick insect? Once you know it is there you can see its body and legs.

Animals with much simpler brains than ours, such as frogs and lizards, probably do not perceive stationary objects as separate things. Their eyes may be just as sensitive as ours, but their simple brains are not as powerful at interpreting what they see. So a fly will be safe until it moves – then the frog spots it and snaps it up!

One thing or another?

Sometimes an image may not contain enough information for your brain to decide which parts of it are in the foreground and which in the background. For example, is the image on page 93 a vase or a pair of faces? You can see it either as a white vase against a dark background, or a

△ Two faces or a vase? You can switch your perception of this image either way.

pair of dark faces against a white background. The picture does not contain enough information for your brain to judge if the background is the dark area or the white area, so you can see it either way.

How far away is it?

Our ability to judge the distance between ourselves and an object can keep us safe and help us to plan our actions. Am I riding my bike too near to the car in front?

Is that angry dog getting too close for comfort? We can judge distances partly because we have two eyes, each of which has an adjustable lens (see page 54). When you look at something nearby you have to focus your eyes. To look at something further away you let your eyes relax. The muscle movements that help to focus your vision on something provide feedback signals to your brain. These signals give you information about the distance to the object.

A 3-D picture

Information about distance is also provided by the so-called 'stereoscopic effect'. Just as a stereo hi-fi system has two separate speakers to create the impression that sound is reaching you from different directions, having two eyes helps to create the impression of depth in your view of the world. Because your eyes are separated, each one has a slightly different viewpoint. Try shutting one eye and then holding up the index finger of each hand so that they appear in line, one behind the other. Now shut the first eye and open the other one. The two fingers seem to move position as your viewpoint changes. The closer your fingers are to your eyes, the bigger the effect.

The stereoscopic effect can fool your brain into thinking that you are looking at a scene with depth, when in fact you are looking at flat pictures. To create a stereoscopic image you have to take a pair of photographs with two separate cameras positioned apart. A special viewer is used to look at one photo with the left eye and the other with the right eye. The slightly different viewpoints produce the perception of depth. Some popular comics and monster movies are made in three-dimensional (3-D) vision. They have to be viewed with special spectacles which usually have a red lens covering one eye and a green lens over the other. The coloured lenses act as filters. When the comic or film is produced, two slightly different images from different viewpoints are printed on top of each other. The colours in the images are adjusted so that one contains more red and one more green. When you look at the two images through the coloured glasses, the different images detected by your eyes combine to create the 3-D illusion.

A new perspective

Stereoscopic vision and eye muscle movements do not explain fully how we are able to judge distance, however. You can prove

▷ Because we have two eyes we have two slightly different viewpoints of our surroundings. Our brains make use of these differences to create our three-dimensional perceptions of the world.

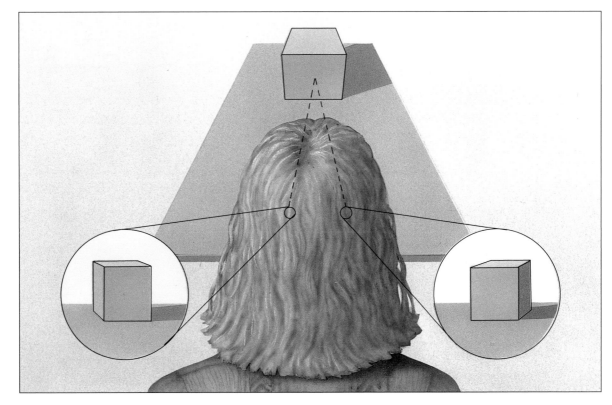

△ Perspective has been used to create a wonderful illusion of depth in this painting completed some time between 1550 and 1603. The distant figures are smaller and higher than those in the foreground. Nearby pillars block our view of those behind. The arches get smaller as they recede in the distance.

this to yourself by shutting one eye – the world does not suddenly turn into a completely flat place. Perception of depth is not as good with one eye, but you still have it. Now you are using your memory and experience to judge distance. Your brain uses your knowledge about the size of the objects it recognizes to judge their distance. The apparent change of size with distance is called 'perspective'.

It was not until the 15th century that artists discovered how to use perspective to create the illusion of depth in a flat painting. Parallel lines appear to meet as they become more distant. An object may be obscured by another in front of it. Distant things are smaller and higher up in the field of view than nearby ones.

In contrast, paintings drawn without perspective appear very flat. The artist

▷ Simple shapes and outlines are instantly recognizable as the objects they represent.

has not understood how to use size and other distance clues to create the illusion of depth. Modern abstract paintings can also appear flat if they do not use recognizable objects or shapes to create perspective.

What is it?

How do we recognize things? When you look at a bicycle you match the object you see to memories of bicycles that you have seen in the past. But what is it about the bicycle that identifies it? Is it its size, colour or some other property? In fact you recognize a bicycle by its shape. We know this because you can recognize a bicycle when it is represented by just two circles and a few straight lines, as in the simple symbol below of a bicycle on a road sign. Our brains are very good at distinguishing different shapes that are built from combinations of curves and straight lines.

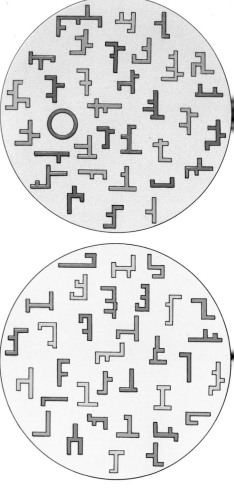

△ The O seems to 'pop out' in the top set. It is the only shape not made from straight lines. You must look at each shape in turn in the bottom set to find the F however, because it is made from straight lines like the other shapes.

Your mind is not only good at recognizing shapes, but it can also group them together to form whole objects. You can arrange a few coins and some matches to make a face or a bunch of flowers. Even if the shapes are not touching, your brain still tries to make recognizable objects by thinking of them as linked together.

Sometimes shape alone is not enough to help you identify an object. A sketch of a sphere could be a football or a planet. You need other clues to decide which it is.

△ B or 13? If you read down the column the context is letters and you see the shape as B. But if you read across the row the context is numbers and you see 13.

These may be provided by the surroundings, or context, in which the sphere is placed. If it is surrounded by players in shorts it is a football. If it is orbiting a star then it is a planet. The context in which you see things helps you to recognize them.

'I'm good at faces'

A large area of your brain seems to be used for recognizing small changes in faces. You can instantly recognize not only all the different members of your family and your friends, but also many TV stars and sportspeople. All their faces have the same basic shapes, yet your brain is incredibly sensitive to tiny differences in the way they are arranged.

Look at your own face in a mirror on a wall. Now lay the mirror on the floor and look again. Does your face look odd? The force of gravity causes tiny changes in the shape of your face as you change from looking at it horizontally to looking at it vertically. You are so sensitive to the structure of your face that even this tiny change is enough to make it look different.

Why are we so good at recognizing faces? Probably because it is so important to distinguish our family and friends from our enemies. We need to be able to detect the slight changes in people's expressions that tell us how they feel and what they are about to do. Is this person about to get angry and attack? Is this person frightened of me? Do I have the upper hand?

△ Top: the mirror is on the wall. Bottom: the mirror is on the floor. Even the small changes produced by gravity as we look down can change our appearance.

▷ One of the most amazing things your mind can do is to pick out a face in a crowd. All faces are composed of the same basic parts, yet you can instantly pick out the face of a friend or a celebrity.

Our awareness of faces can make us see faces everywhere. We can see a face in the front of a car, in a wallpaper pattern or in the clouds. Perhaps the face-making ability of our brains is responsible for the way some people claim they have seen spooks and ghosts. Our minds change unfamiliar shapes and shadows into faces.

Paying attention

Your perception of sounds has a lot in common with your ability to see things. Just as your eyes allow you to see, your ears help you to judge the distance and direction of sounds. You also use your knowledge and memories of sounds to locate and recognize them. For example, you know that a buzzing insect must be close by because it cannot make much sound. The muffled sound of a plane high up in the sky, on the other hand, must be in the distance because it is not very loud.

We can choose which sounds, and other signals, to pay attention to. Inputs from all your five senses arrive constantly in the brain. At a noisy party, for example, the lights are flashing, the music is loud and you are being jostled by other dancers. Yet you can usually filter out all these signals to pay attention to a friend's whispered conversation. But you do not block out the other inputs completely. If your favourite record starts playing you will probably start listening to it.

Try switching your attention for a few seconds. Stop reading these words and listen to the sounds in the room. How many different ones can you hear? Where are they coming from?

Of course you do not always choose where to direct your attention. It can also be focused by a very strong sensation. Warning lights, sirens and the smell of burning are good examples: their input is so strong that it forces your mind to pay attention to their message. Pain grabs your attention too – if your socks caught fire you would stop reading immediately!

Knowledge and expectations

Your perceptions – your awareness and understanding of the signals from your senses – result partly from the nature of the signals themselves. Are the shapes straight or curved? Are the sounds loud or soft? Perceptions also result from your knowledge and expectations of the world around you. An antique dealer can recognize a fake oil painting from a genuine one. The differences between the two paintings are there for everyone to see, if only we knew where to look. Knowledge and experience draw the dealer's attention to those differences, so that he or she can perceive them.

Read the words in the triangle below. Do they say ONCE UPON A TIME? Actually they say ONCE UPON A A TIME. You probably missed the extra A because once you started reading you expected the familiar phrase.

Our perceptions can depend as much on what we expect as on what we actually sense. When you see the Queen on television you know her immediately. Yet if you

△ When viewed from an angle the circular face of a coin looks as if it has been squashed into an oval.

bumped into her in your local supermarket you probably would not recognize her because you do not expect to see her there.

Same or different?

We also expect that things will usually stay the same shape and colour when they move from one place to another. When a circular coin is viewed from different angles its image varies, from a perfect circle to a straight line. Yet we always perceive the coin as circular because we expect its shape to be constant.

Colours, on the other hand, can be deceiving. The colour you see depends on the amount of light that is reflected from an object. Photography proves that colours change in different lights. If a bowl of fruit is photographed in sunlight and in artificial light, the fruit will be different colours in the two photos. Yet you do not see the difference when you inspect the fruit with your eyes. You expect bananas to be bright yellow so this is the colour that you see when you look at them in both sunlight and artificial light. Your brain adjusts its responses to the signal it receives from your eyes, so that you see the colour you expect.

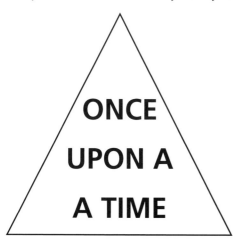

GETTING IT WRONG

Perception is not just a matter of sensing what is there. It also involves making sense of the information. Sometimes we get it wrong. Illusions are a fun way to show some of the ways in which our perceptions can be mistaken.

Which line is longer?

Which of these lines is longer – the upper or the lower one? In fact they are exactly the same length. The upper one looks longer because the converging rails make you think you are looking at a three-dimensional scene. In three dimensions the more distant line would be longer. Here your perception of perspective fools you.

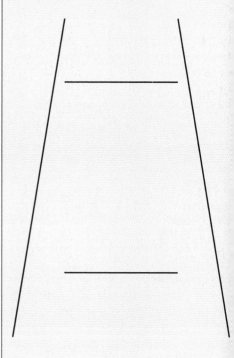

The Ames room

Is this a trick photograph (right)? No, it is a trick room! The girls are identical twins and they are standing in the room at the same time. The trick works because you expect a room to have rectangular walls and windows. In fact the walls and windows of this room are not straight, and the left corner is much further away than the right one. Here you are tricked by your knowledge and expectations of the size and shape of rooms.

Impossible shapes

Imagine you are one of the figures in this painting by Escher. You are doomed to climb or descend for ever, never reaching the top or bottom of the impossible stairs. At first glance Escher's building looks quite normal. Your mind expects the world to make sense. You are used to using clues in a two-dimensional drawing to see a three-dimensional object. But on looking closer, you can see that the building could not possibly exist in three dimensions. Escher cleverly distorts perspective and introduces contradictory visual clues to create fascinating images of impossible buildings and shapes.

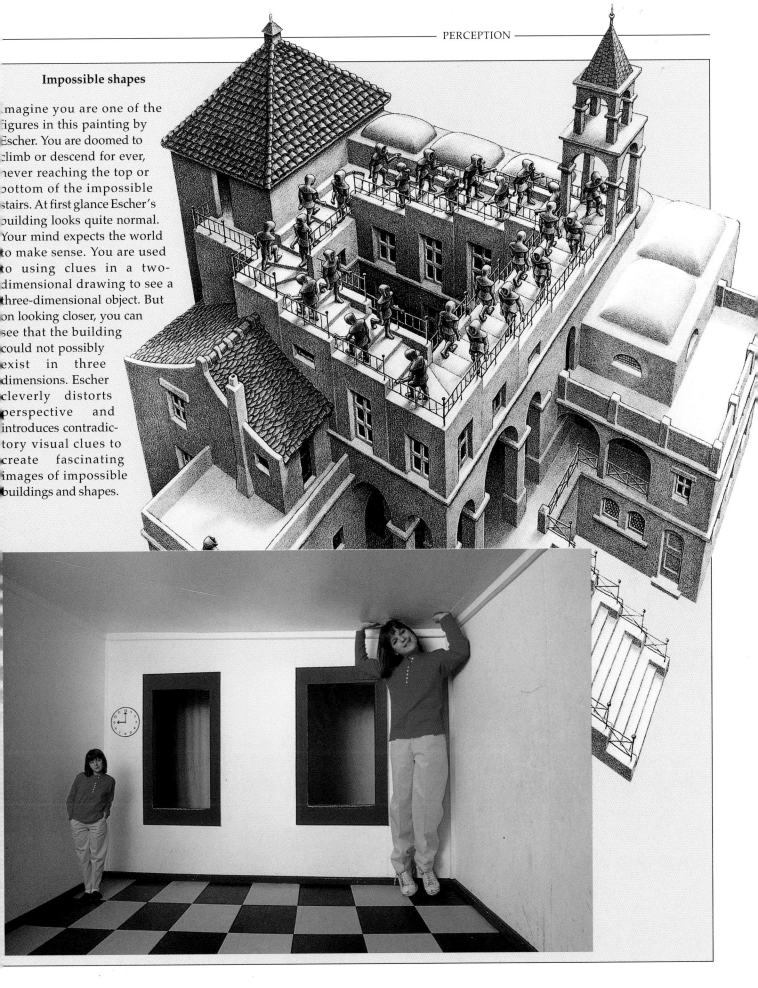

LEARNING AND REMEMBERING

What is your first memory? Some memories stay with us our whole life, but others are gone in a flash. Memories are patterns in our brain, and new patterns are created when we learn. Understanding how our memory works can help us to learn more effectively.

Your memory is phenomenal. You can remember sights, sounds, smells, tastes, textures and movements. You have the capacity to sort through faces in a crowd and find people you recognize incredibly quickly. You can instantly identify a voice on the phone from hundreds of people you know. You probably understand the meaning of thousands of words and can recall countless experiences from your life. Your ability to do this is far greater than that of any computer, and you will still have much of this information stored in your brain in 60 years time.

But we do not automatically remember everything we see and do. Telephone numbers are a good example. How many times have you started to dial a number that someone has just told you and forgotten it in the middle of dialling because you were interrupted? This is an example of 'short-term' memory. Most people cannot remember a series of six or more digits for more than a few seconds. You can, however, commit a telephone number to your 'long-term' memory by writing it down and repeating it to yourself, or using it frequently. How many telephone numbers do you have in your long-term memory?

What are memories?

We still do not fully understand how memory works. Research shows that a single memory is not stored in one place in the brain like a card in a filing cabinet or data on a computer disk. Each memory probably forms a pattern of links between the billions of nerve cells that make up your brain (see page 86).

The first time you experience something, for example drinking coffee, flying in a plane or hearing a song, a pattern of nerve impulses is created in the brain. In the case of short-term memory, it is possible that the pattern soon fades, and so the experience is forgotten. But if the experience is strong, or is repeated several times, then the new pattern is reinforced. It is then transferred to a part of the brain where it becomes permanent, and a long-term memory is formed. When you remember something, the pattern of brain impulses produced by the original experience is recreated. The fact that your brain has experienced this pattern before causes you to remember, although we do not know exactly how.

Picturing memories as patterns in the brain helps us to understand some of the ways in which our memories work. If each memory was placed in a separate box or compartment, then we would expect every memory to be separate, but if memories are patterns that spread and overlap, we can understand how they are often linked or associated. For example, when I listened to the first pop record I ever bought, my mother was cooking bacon in the kitchen. Now, whenever I hear that song I recall the smell of bacon. People sometimes have personal

◁ The human being has the unique capacity to pass on the learning of a lifetime through language and literature.

memories associated with a dramatic world event. These are called 'flashbulb' memories. Most Americans over the age of 20 can remember exactly what they were doing when they heard that the space shuttle 'Challenger' exploded on 28 January 1986.

Very often we remember the things that are most important to us. This is known as 'selective' memory. For example, you may easily remember the title of the new hit single of your favourite group, which you have just heard on the radio, but you may have more difficulty remembering that mathematical equation! Perhaps you find it easy to remember the names of the players in your favourite football team because you find their successes and failures so exciting.

> Through constant repetition we develop 'muscle memories' and learn to perform difficult tasks with apparent ease.

△ Remembering names can be made easier by making mental images. Recalling the pictures above brings the names Francis Cook and Robin Waterman to mind.

Improving your memory

Can you improve your memory? You can learn and remember more effectively if you link new ideas to strong visual images or to things that you already know. When you associate two facts or ideas in this way, recalling one usually brings the other to mind. A trick to help you remember a name you have forgotten is to go through the letters of the alphabet, trying out the sounds of different letters until you find the one that fits. Suddenly the association between the sound, the name and the person will trigger your memory.

Mnemonics are tricks for remembering facts. 'Richard Of York Gave Battle In Vain' is a mnemonic for remembering the order of the colours of the rainbow – red, orange, yellow, green, blue, indigo and violet. This order is not particularly meaningful on its own, but by associating the colours with the initial letters of a sentence with meaning, the rainbow colours are more easily recalled.

What everyone knows

You probably remember learning to ride a bicycle or going to swimming lessons at the pool. Mastering these skills takes practice, patience and effort. So why can other animals do some things so easily? Spiders can spin beautiful webs without any lessons. Some birds can fly as soon as they leave the nest. Ants build their nests without being shown what to do.

Most of these incredible skills possessed by animals are 'innate' – they are programmed by genes (see page 13). A spider does not have to be shown how to make a web to catch its prey.

Human beings are born with some innate capacities. Every child learns to walk and talk when it is ready. But children do not learn to write unless someone teaches them and encourages them to keep trying. Writing, along with most of the things we learn, is an 'acquired' skill. The ability to acquire and invent a great variety of new skills marks human beings out from other animals. We do not have the innate ability to fly but, by applying the human mind to the problem, we have created aircraft that can fly higher and faster than any bird.

Simple learning

Some famous experiments on simple learning patterns were carried out at the start of the 20th century by a Russian scientist called Pavlov. Using dogs, he showed that the brain can learn to associate two events if one always follows the other. This kind of learning is called 'conditioning'.

In childhood some of our learning is conditioned with rewards and punishments. We learn to associate certain behaviours with certain outcomes. A young child may soon learn that if she messes around with her food at mealtimes, it may be taken away, whereas if she eats properly she may be rewarded with her favourite dessert. Some scientists have tried to develop educational strategies based on this simple idea of reward. It has

▷ Motivations for learning: survival, a rewarding career or simply the pleasure gained from mastering new skills.

been shown that pupils learn better when they are encouraged and praised by their teachers, than they do if their work is constantly criticized. Computers are now able to reward a student with praise every time a correct answer is given to a question. But we soon get bored with learning things just for the sake of a message saying 'Well done!'. We need better reasons to learn than this.

Motivation

We can be motivated to learn for a variety of reasons. Imagine that you are stranded on a desert island. You must learn how to find food and make a shelter or you will die. Your motivation for learning is straightforward – survival. At the other end of the scale we learn some skills purely for pleasure. In developing hobbies, for example, we are motivated to learn for the enjoyment and interest we gain. Whatever the motivation, having a reason for learning makes it more interesting and enjoyable. Learning is more effective when we can see the benefit.

◁ Pavlov rang a bell before feeding his dogs. The dogs soon began to salivate on hearing the bell. They had learned to associate the bell with the food that followed.

IMPROVING YOUR LEARNING

Studying science, mathematics or literature is more involved than learning a physical skill such as juggling. There are facts to be memorized, new terms to be understood, problems to be solved, practical skills to acquire, patterns to be recognized and links between ideas to be made. But you can improve your ability to learn by applying some of the same learning techniques and by approaching each new learning task in the appropriate way.

The key to learning a new subject is organization. You must organize the information and ideas you need, in ways that you can understand and recall. Just reading a book from cover to cover probably is not much help in doing this. Note-taking, highlighting key words, attempting to summarize what you have just read, solving problems, writing essays and having discussions with fellow students are all effective ways to help you structure new knowledge in your mind.

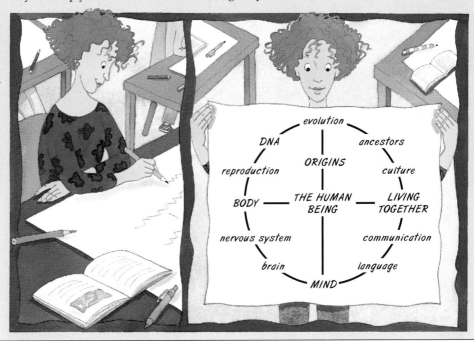

△ To learn to juggle you need to watch an expert, analyse their actions and, through trial and error and constant practice, refine your own skills.

▷ By organizing ideas on a 'mind map' you will discover patterns and make connections that improve your understanding of a subject.

Learning through life

A newborn baby has so much to learn. The first years of our lives are filled with new sights, sounds and experiences, and it is during this period that we learn most rapidly. Stimulation by adults, through talk and play, is essential if a child's mind is to develop its full potential.

As a child develops, she begins to understand the things around her and to experiment with the meaning of words. A baby will point to picture of a dog in a book and say 'doggie'. She may then point to a real dog in the street and say the same word. The picture in the book, the sound of the word, and the four-legged animal have become associated in her mind. She has learned that they mean the same thing. At this stage the child will often make mistakes, perhaps saying 'doggie' when she sees a cat. When an adult corrects her by saying 'No, cat', then she learns the new word and will soon be distinguishing cats from dogs.

Perhaps it is only as we come to understand the meaning of the different things in the world – dogs, people, houses, grass, cars and so on – that we become aware of ourselves as individual people, separated from our surroundings. This may explain why we have no conscious memories of the first two or three years of our lives – at that stage our minds have not yet understood the world in a way that we can remember. But when we start to give things meaning, we acquire knowledge, and memories, at a tremendous rate. By the age of 18 an intelligent young adult might know the meaning of 50,000 words. This means that, on average, he or she has learned eight new words every day of their life.

Learning does not stop when we leave school. Many people keep their minds active throughout their lives into old age. New skills and knowledge are acquired through work, raising a family and pursuing our interests. Some older people who return to study complain, however, that they find it harder to remember things than they did when they were young.

It is true that the elderly sometimes have poor memories for recent events. This suggests that the parts of the brain that form memories are no longer working as efficiently as they once did. It is not uncommon to meet an old person who has difficulty in remembering what they did the day before, but can recall details from their childhood 'as if it were yesterday'.

INTELLIGENCE AND CREATIVITY

All human beings are intelligent when compared to an ant or even to a chimpanzee. But what makes a particular person a genius? Can you improve your intelligence, or are you born with it?

There is no clear definition of intelligence. Is an absent-minded scientist who invents theories about the Big Bang, but who cannot boil an egg or wash a pair of socks, more intelligent than a hunter-gatherer who can navigate by the stars, light a fire with two dry sticks and find water in the desert? Each one has a kind of intelligence that the other lacks.

In the modern world we sometimes associate intelligence with knowledge. We admire people who know the answers to all the questions in a quiz, and often tease them by calling them 'swots'. The ability to absorb information and memorize details of experiences certainly has something to do with intelligence. But remembering facts does not make a person intelligent. After all, a computer has a much more accurate memory than a human being, and we do not think of computers as being intelligent.

The people who we really think are intelligent are those who use their knowledge to produce new ideas, to find the solutions to problems and to create new ways of understanding our world. In this sense intelligence takes many forms. Some people use words well and can express their ideas in prose or poetry. Some people are skilled with numbers – they can make accurate calculations, understand equations and interpret the meaning of statistics. Some people have a good visual intelligence, for example an architect who can see how to design a striking building. Some people's intelligence is concerned with dealing with other people. They might be teachers who are gifted at explaining things, comedians who know how to make us laugh or politicians who understand how to influence other people's ideas. Some people are mechanical geniuses who can repair anything that is broken; others are gifted at music, dancing, gardening or woodwork. The number of ways that we can show our intelligence is almost unlimited.

Nature or nurture?

Particular talents or interests are sometimes found in members of the same family. The children of musicians, for example, often become musicians themselves. Sportspeople often have sporting children; the sons and daughters of doctors often study medicine when they grow up. Does this prove that we inherit our talents from our parents, through their genes? Not necessarily. It could be that because a musician's child hears a great deal of music at a very young age, and is encouraged by enthusiastic parents to make music, he or she develops musical skills as a result of these childhood experiences.

People argue at length about the extent to which our intelligence and talents are inherited in our genes, or learned through our experiences. Many studies have been undertaken to try to establish whether it is our nature (our genes) or our nurture (our upbringing) that determines how intelligent we are. Some of these studies have created tremendous controversy as it is very difficult to decide whether an influence is a genetic or an environmental one. The example of the musicians' children

◁ Does going to school make children more intelligent? We need the encouragement and guidance of good teachers to make the most of the natural talents we all have.

Whatever genes you have, you will not become a great scientist or a successful musician without many hours of dedicated study. Perhaps more important than any 'natural intelligence' you show for a particular activity is your motivation. The more you want to do something the more time you will spend on it, and so the better you will become at your chosen activity. People who are really successful in a particular field are not necessarily those with the greatest natural ability, but rather those who work hardest at it. It is no good being a natural dancer or footballer unless you train to perfect your skills and make use of them.

demonstrates this difficulty. This debate is known as the 'nature versus nurture' debate.

One possible approach to answering the 'nature or nurture' question is to study identical twins who are brought up in separate homes. We know that the twins' genes are the same, so the differences that develop between them should be explained by their different upbringing and environment. Such twins are quite rare.

The few twin studies that have been made suggest that both nature and nurture are important. We probably inherit certain skills or potential talents from our parents, but the way our talents actually develop depends on how they are encouraged by our parents and others around us, and at school. This is an important point for our education, since it means that no child can be dismissed as simply unintelligent or lacking talent. Even adults who have always thought of themselves as completely tone-deaf can learn to sing and appreciate music, when they are taught in an appropriate way.

△ Intelligence takes many forms. Kinesthetic intelligence is the ability to control your body and other objects skilfully. Racing drivers, pilots and competition skiers must develop this intelligence to a high degree.

▷ This sculptor has highly developed spatial intelligence. He can imagine shapes in three dimensions.

A vivid imagination

The human ability to create and invent has much to do with our imagination. Somehow our minds have the capacity to experiment with what we know about the world and to create mental pictures of things that have never existed outside our minds. We can imagine our future lives. We can write novels and plays. We can even create fantasy worlds filled with impossible creatures and events.

Scientists and engineers also rely on their imagination to create new theories and structures that have never existed before. Their imaginings are different from those of a novelist or poet, however, because they must be tested in the real world. A scientist carries out certain experiments to find out if a new theory is false. The final test of a new bridge imagined by an engineer is whether or not it falls down when it is built.

Why do we have this capacity for imagination? Humans have always lived in groups, and it has been suggested that our imagination evolved so that we could try to predict the behaviour of others. If you can imagine what someone else is thinking, or what another person might do in response to your actions, this gives you a considerable advantage. Is that person about to attack? Will he or she back off if I am aggressive, or should I run and hide? In this situation an intelligent person is someone whose imagination takes account of all the clues and evidence from their senses and previous experience to predict the future, and is therefore able to make the most sensible decisions. Today we can make good use of this imagination to create everything from cartoon characters to great works of art and exciting scientific theories.

Creative thinking

'It came to me in a flash.' This is a popular expression used to describe a scientific breakthrough, or the composition of a song. The final creative idea seems to come 'out of the blue' – it is inspiration from nowhere. People then have the mistaken

◁ The dinosaurs were extinct millions of years before humans evolved, but the human mind is able to imagine what it might be like to meet a dinosaur.

idea that you can suddenly discover a new scientific formula or write a chart-topping song after just a few days of effort. This is misleading. People who create original ideas have normally spent a tremendous amount of time thinking about their subject. They have studied what others have done, experimented with new ways of looking at things, and perhaps have experienced many failures before they achieve success. People say that genius is 99 per cent perspiration and 1 per cent inspiration!

A famous story about a sudden inspiration is often told about the German chemist Kekulé. He knew that the chemical called benzene contained six carbon atoms and six hydrogen atoms, but he could not see how these atoms could link together (and still follow the laws of chemistry). He had tried many different combinations but none of them was correct. Then one day on a bus journey he was daydreaming. In his mind he pictured a snake biting its own tail. Suddenly he realized that this was the shape he had been searching for. The atoms in benzene are arranged in a ring. Of course Kekulé would not have had this inspiration if his mind had not been filled with the details of the problem.

Like Kekulé's snake, creative thinking often takes place as pictures rather than words or numbers. Some scientists suggest that this is because our creativity is located in the non-verbal right hemisphere of our brain rather than in the verbal left hemisphere (see page 90). Another example of visual thinking in science is the idea of a 'magnetic field' created by Michael Faraday. Many scientists consider that Faraday, a blacksmith's son, is the greatest experimenter of all time. He was not a gifted mathematician, however, and preferred to think in pictures rather than in mathematical equations. To picture the force created when one magnet is placed near another, he imagined a pattern of 'lines of force' spreading in the space around the magnets. This incredibly powerful idea is now used throughout science whenever a force field is discussed.

Solving problems

Human intelligence is needed to solve the classic problem about the farmer who has to cross a river with his dog, chicken and corn. He can only carry one of the three things across the river at a time. The farmer's problem is that if he leaves the chicken with the corn, the chicken will eat the corn. If he leaves the dog with the chicken, the dog will eat the chicken. How does he get all three things across the river safely?

How does your brain go about solving a problem like this? If you know the starting-point, and the end-point, then the way to solve a problem is to break it down into a series of steps which add towards the correct outcome.

▷ The farmer's dilemma. How can he get his dog, chicken and corn safely across the river? To solve this famous problem you must think at least two steps ahead.

Obviously the only item the farmer can take across the river on his first trip is the chicken, leaving the corn safely with the dog. But what about the next trip? If he takes the corn, the chicken will eat it when he returns for the dog. If he takes the dog, it will eat the chicken when he returns for the corn!

The solution to the farmer's dilemma is that he must take the corn next, but then he needs to bring the chicken back across the river when he returns to pick up the dog. Now he can leave the chicken alone, take the dog across and leave it with the corn. Finally he returns to collect the chicken.

Choosing a strategy

We must tackle all kinds of problem in life. A few, such as mathematical problems in a textbook, have definite answers. They can sometimes be solved by working backwards, and this is how many beginners in a subject tackle a problem. If you look up the answer you can try and work back, a step at a time, to the original problem.

However, you have to work forwards in order to solve the more realistic problems that you meet in everyday life. You cannot look in the back of a book to find the answer to problems such as 'How do I get home now that I've missed the bus?' Most problems have more than one possible solution. Having missed the bus you could, for example, find a taxi, ring home and ask your mother to fetch you, borrow a friend's bicycle or start to walk! A good strategy for solving most real problems is to imagine what a satisfactory solution would be like, and then formulate a plan to try to achieve it.

Tests and exams

We take tests and examinations throughout our lives. Some examinations test our potential or ability to learn before we start a new course or job; others test what we have already learned or achieved during a programme of study. New recruits to the army, for example, are given tests to see if they are best suited to train as engineers, clerks or officers. A driving test checks that you have learned all the skills needed to be a safe driver.

In an IQ test you have to answer questions that test your verbal, numerical and

△ Albert Einstein is widely regarded as the most creative scientific genius of the 20th century. His theories of relativity have revolutionized our understanding of space, time and the origin of the Universe. Yet as a young man he failed his university examinations.

visual reasoning and short-term memory. The result is a single number which is supposed to show how 'intelligent' you are. An IQ score of 70 is classed as 'retarded', 100 is 'average', 120–130 'superior', and 130 or more means 'very superior'. But these tests have been criticized for a number of reasons. For example, they take little account of the many different forms that intelligence takes. Someone who is a skilled mechanic or who plays a musical instrument well might perform badly because they are not good at coping with words and numbers. Also, IQ tests take no account of a candidate's cultural background; some children might take the test in their second language.

Most significantly, there seems to be little connection between the way that people perform in IQ tests and what they achieve in later life. Some of the greatest contributors to science and the arts have performed only moderately well in IQ tests and exams. Other people with high IQs never manage to achieve anything of note. Some people say that the only thing an IQ test really tests is the ability to do IQ tests!

TEST YOUR IQ

1 Spatial/visual

Which one of these cubes could <u>not</u> be made by folding this shape?

a)   b)   c)   d)

2 Numerical

What is the missing number?

2	3	5
4	5	9
3	4	?

a) 5   b) 7   c) 9   d) 11

3 Verbal

Which pair of words expresses a similar relationship to GOOD–BAD?

a) BEAUTIFUL – ATTRACTIVE
b) RED – YELLOW
c) SMALL – TINY
d) CLEAN – DIRTY

Answers: 1:d 2:b 3:d

A modern approach to testing in schools is to take full account of a pupil's efforts and achievements over an extended period of time. Pupils are not tested in a single examination in which they might not do particularly well. Some pupils achieve consistently good marks during the school year but then perform badly in an exam. A portfolio which records pupils' achievements throughout their school career gives a much more rounded picture of their motivation and abilities.

How do I become a genius?

Geniuses are people who make exceptional new discoveries or creations. Not all geniuses were outstanding youngsters. Einstein, for example, failed his exams at university and took a fairly ordinary job as a clerk before creating his theory of relativity. Geniuses are, however, totally absorbed by their subject. They spend many more hours working at it than the average person, and this extra effort is reflected in their achievements. It is not necessary to be exceptionally talented to become a genius – after all, many talented people do not use their abilities fully – but you do have to be exceptionally motivated.

△ Mozart worked quickly, writing down his musical compositions with very few corrections. This is part of his original score for his Symphony No. 40.

▷ By the age of six Mozart was an accomplished musician. With his father and sister he toured the royal courts of Europe.

FROM CHILD PRODIGY TO GENIUS

Sometimes children possess skills and talents that you would expect to find only in older, more experienced people. These are child prodigies. Many have been pushed forwards early in their lives by ambitious parents, and often their talents do not develop beyond their teenage years. In rare cases a child prodigy does develop into a true genius. The composer Wolfgang Amadeus Mozart is an outstanding example. He came from a highly musical family and was composing at the age of six. By his death at the age of 34 he had written hundreds of musical masterpieces.

ALL IN THE MIND

How do you know if you are awake or dreaming? Can you control your thoughts and feelings, or do they just happen to you? Questions about our minds are fascinating, but they are some of the most difficult questions about human beings to answer.

Since the time of the very earliest humans, people have believed that every human being has two separate parts. Your body is the physical part, and your mind is the mental part. Your mind does not belong to the physical world – you cannot see or touch it. It has several different names, including 'spirit' and 'soul'. Your mind, or spirit, is the important part of you that determines your personality. It feels emotions such as love, pity and anger. It can be kind or cruel, and can choose between good and evil. The belief that each person has a spirit or soul that lives on in some way after death is part of many of the world's great religions (see page 132).

Mind and matter

If your mind is separate from your body, then where is it? How can something which is not part of the physical world influence how your body behaves? For example, when you think about a good friend, kind thoughts produce the nerve impulses that may make your face smile. Yet if your mind is not part of your body, how can this happen? The two must be connected in some way. Philosophers have discussed this mind–body problem for centuries.

In the 20th century scientists have made many new discoveries about the incredibly complex structure of the human brain. Many now think that the human mind is really just the activities of the billions of neurons (nerve cells) in our brains. They think that everything about us, including our personalities, our feelings and our knowledge of ourselves, can be explained by the genes we inherit and by the way that our brains and bodies develop as we grow up. But are they right?

Is there anyone out there?

The 17th-century philosopher René Descartes said: 'I think therefore I am'. You know that you exist. But is everything else in the world real? Do you sometimes wake up from a dream and realize that the people and places you have just been visiting were not real – they existed only in

▽ The human mind has the amazing ability to divide its consciousness. You can concentrate on riding a bike, talking to a friend and planning when to do your homework, all at the same time.

your mind as you slept? Perhaps the whole world, even your own body, is just a dream! We can never prove that this is not so, but most of the time we accept that the world outside our minds does exist. We do this because the real world seems to make sense. The things that happen in our dreams are often strange, different from the things that happen in real life.

You know that other people exist, but how do you know that their experience of the world is like yours? For example, when you look at a red tulip, how do you know that your friend is seeing the same thing when she says that the tulip is red. Perhaps the colour she sees as red looks like green to you. We can never experience someone else's feelings or sensations directly. Yet when we talk about our experiences, we can usually agree on the nature of red and its different shades. In the same way, we usually agree when something smells bad or tastes sweet. This gives us confidence that other people do experience the world in the same way as we do.

Conscious or unconscious?

The thoughts that we know about and pay attention to are called 'conscious thoughts'. When you watch a film you are conscious of the action on the screen and of the emotions, such as sadness or fear, that the film produces in you. But your brain does not completely switch off from everything else that is happening around you. As well as listening to the soundtrack of the film, your ears detect other sounds in

the cinema. Perhaps the people in the row behind you are talking. While you are watching the film you are not conscious of what the people behind are saying. But if you suddenly hear your name mentioned, you might switch your attention from the film to what they are saying. Although you were not consciously listening, your brain

was subconsciously monitoring the conversation. Because your brain can keep track of several things at once, you can switch your attention from what you are doing, for example watching television at home, to becoming suddenly aware of a danger signal, such as the smell of burning in the kitchen.

△ Both these boys are clearly enjoying the same joke. We usually agree when something is sad, frightening or funny. Different human beings seem to experience the world in similar ways.

▷ A soldier upset by the terrible events that he witnesses while on duty may keep an impression of them in his unconscious mind for a long time.

SEEING SOUNDS

Some people have unusual experiences that we cannot share and are difficult for us to imagine. For example, when some people hear a sound, they 'see' it as a colour. This ability is called 'synaesthesia'. Most people with synaesthesia agree that the vowel sound 'i' is coloured white to pale grey, 'o' is white and 'u' is yellow to light brown. Some synaesthetics see smells as colours or hear a sound when they see a certain colour. These people may have extra links in their brain between the areas that perceive colours, sounds and smells.

At the beginning of the 20th century the Austrian doctor Sigmund Freud suggested that, as well as conscious thoughts and feelings, people also have unconscious feelings, memories and desires. These are thoughts that we have hidden, or suppressed, from our consciousness (the thoughts that we are aware of). We may do this because we are ashamed of them, or because they are very unpleasant thoughts. Freud thought that the unconscious part of our mind could shape our personalities. He also thought that the unconscious could cause mental problems. Freud developed a technique called 'psychoanalysis', in which patients could reveal all their unconscious thoughts by talking to a trained analyst. Freud's ideas are very controversial, but they have had a strong influence on the way we think about the human mind.

You are unique

Like every other person, you have your own unique personality. It is made up of all your conscious and unconscious thoughts and feelings, and the way you present yourself to others, for example as a confident or a shy person. In our minds we can sometimes be anything we choose – a poet, a pop star or a politician. But the person we actually become is more complex and more varied than any fantasy character. At different times you can be thoughtful like the poet, outgoing like the pop star or assertive like the politician.

The ancient Greek thinker Hippocrates suggested that there are four personality types, and that all human beings belong to just one of these: melancholic (depressed), choleric (irritable), sanguine (optimistic) and phlegmatic (calm). You probably know people who fit into each of these types. But there are more than just four kinds of people in the world. Some modern psychologists prefer to see our personalities as a complex mixture of different features. Are you calm or do you worry a lot? Are you shy or sociable? Are you unadventurous or daring? Are you self-centred or kind? Are you unreliable or conscientious?

If you visit a careers adviser you might be asked to take a test to find out about your personality. This helps the adviser to match your personality to suitable jobs.

△ Are you adventurous, outgoing, thoughtful or shy? Do you enjoy your own company or prefer to be in a crowd? Your personality characteristics will affect your choice of work and leisure activities.

▽ Each of us has a mixture of different personality features. If you are warm-hearted and patient you may choose to work in a caring profession.

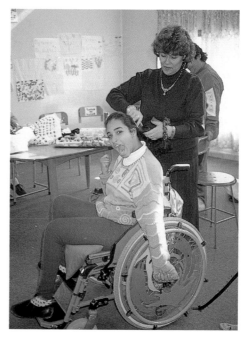

You might be asked a question such as 'Do you prefer working on your own or as part of a group?' Your answer would help the adviser to establish if your personality is better suited to working with people or with things. This type of test can help you think about what is involved in various jobs. But because everyone's personality is different, there is no guarantee that the test will produce the ideal job for you.

Sleep and dreams

Sleep is a time of rest for your body, but not necessarily for your mind. At various times during the night your mind is active. This is when you are dreaming. Some people claim never to dream. Do you ever wake up and feel that you have slept well

△ Some people say that if you sleep with a piece of wedding cake under your pillow you will see the man you are going to marry in your dreams.

WHILE YOU SLEEP

Scientists can monitor the brain activity of a sleeping person by attaching electrodes (electrical connections) to their skull. The electrodes are connected to an instrument with pens and a long chart of moving paper. The electrodes pick up the electrical activity of the neurons in the brain and record this activity as a wiggly line on the chart. This chart is called an 'electroencephalogram', or EEG. When you are awake your brain is very active and the pen jiggles about rapidly on the EEG. When you fall asleep the EEG gradually becomes smoother as your brain activity slows down, and you fall into a deep sleep. About an hour after you fall asleep, the EEG shows that your brain is active again. You are still sleeping but things are happening in your mind. At the same time your muscles become completely relaxed, apart from those that control your eyes. Beneath your closed eyelids your eyes are moving around rapidly as if they are following the events in your mind. This is called 'rapid eye movement' (REM) sleep. If you wake up during REM sleep you will nearly always remember that you have been dreaming. REM sleep can last for many minutes. During the night you may pass in and out of REM sleep several times.

---

all night with no dreams? We probably all dream every night, but it is only when we wake up during a dream that we remember we have been dreaming. Even then, if you do not make a special effort to remember a dream, it is soon forgotten.

What are dreams? Over the centuries, people have tried to find deep meanings in people's dreams, or to use them to predict the future. Sigmund Freud thought that people's dreams revealed their unconscious wishes or fears. Your dreams are often connected in some way to recent events or feelings in your life. Yet the people and events in the dreams become muddled with memories from your past, and with images from films or television. The events in our dreams are often completely nonsensical. If I am worried about something, I often dream that I am about to take a school exam, even though it is many years since I last took one!

Today, some scientists think that our dreams are produced as our brains are tidying themselves up. Our brain may tidy up the various connections and links inside it, as it sorts out our experiences of the day and files away our new memories. Perhaps the brain activity created by a recent event or worry is similar to some other experience that your brain has stored as a memory. The two may become confused as your brain sorts them out. This might explain my dreams about school exams. One function of the activity of your brain during dreams may be to get rid of unwanted information. If this is the case, then it might be better to try to forget your dreams after all!

MIND AND HEALTH

A healthy mind in a healthy body is alert and interested in the world. It absorbs information, produces ideas, copes with problems and runs your body smoothly. But, like any body system, minds can sometimes have problems.

Your mind is created by the activity of your brain as it processes information from your senses and produces your thoughts and feelings. On most days you probably feel happy and enthusiastic and everything seems to be going well, but do you sometimes have 'off' days when you feel tired and bored, miserable and depressed? Everyone has days like these. Usually we soon recover from such negative thoughts and feel bright and alert again. But sometimes depression, anxiety (worry) or other mental problems, such as fear or confusion, can take people over and prevent them from leading a normal life. In severe cases these problems are described as 'mental illnesses'.

Doctors are still not clear about what causes every kind of mental problem. In some cases these problems are due to an infection or damage to the brain's structure; for example, if you have a blow on the head it can cause loss of memory. In other cases the problem could be connected to the information that our minds have stored, particularly from early experiences in childhood. A child who was attacked by a dog, for example, may grow into an adult with a phobia (an exceptional fear) of all dogs. This phobia may be so strong that the person will not go out of the house in case he or she comes across a dog. The person is unable to lead a normal life because of the phobia.

A delicate machine

Your brain is the only organ in your body that is completely enclosed in a bony case – your skull. The presence of this bony shell reflects just how delicate our brains are and how important it is to protect them. Unlike other parts of your body, brain tissue does not repair itself when damaged. If you suffer a blow to the head, the brain twists and wobbles like a blancmange inside your skull, tearing the brain tissue and blood vessels. This causes internal bleeding which produces pressure that further damages the brain. Many people are concerned about the safety of sports such as boxing, which involve repeated blows to the head. The damage which this causes may create mental problems for boxers later in their life.

We know that some mental problems are caused by diseases that damage brain tissue. Alzheimer's disease, for example, which affects some elderly people, makes many brain cells die. Sufferers become confused and forgetful, and eventually they are incapable of looking after themselves. The cause of this disease is still not clear.

▽ Healthy minds are active and alert, ready to be stimulated by, and to respond to, excitement and new challenges.

◁ Repeated blows to the head damage the brain's delicate tissues. Some doctors think that professional boxing is too risky and should be banned.

Feeling moody

Our moods can change from one day to the next, and even from one hour to the next. You may wake up feeling tired and miserable, but after eating breakfast and seeing sunny weather, you may feel happier. This is quite normal. Mood changes are often triggered by external events. Before school exams you may feel anxious and irritable as you are busy revising. An unexpected gift or an invitation to a party can put you in a good mood. The death of your favourite pet can make you very sad. Some moods happen for no apparent reason – you just feel bright and happy or dull and sad on a particular day. Teenagers have a reputation for being moody, and it is probable that their mood swings (changes) are associated with the changes happening to their bodies and their emotions. These changes are produced by the hormones released during puberty (see page 59).

Raising your spirits

What can you do to improve your mood when you feel miserable? Some people become reflective when they are moody. They think about themselves and their feelings. They may find that a solitary walk in the countryside or expressing their feelings in poetry helps them to have a more positive outlook, but this is not always the case. In general, moods seem to last longer for the reflective person than for someone who takes more positive steps to alter their mood.

If you can find out what is causing your mood, you might be able to do something to influence it. When your cat dies, buying a new kitten may help to cheer you up. This is not a heartless act; you still feel sad about the loss of your old cat, but the responsibility and fun of having a new kitten will put your feelings into perspective. Similarly, the best way to cope with the worry of exams is to involve yourself as much as possible in the preparation and revision for them. If you organize your notes and practise answering questions you will not have time to think about your fear of failing.

Physical activity is now recognized as having a very beneficial effect on our moods and sense of wellbeing. Negative thoughts will often vanish if you get involved in an energetic game of tennis, an aerobic work-out or even digging the garden. As our bodies become fitter and healthier through exercise, we also feel better about ourselves. Exercise seems to encourage the release of various chemicals, called 'neurotransmitters', in our brains (see page 87). These make us feel brighter and more positive about things. Some kinds of depression may be associated with a lack of these chemicals, although it is not clear whether this shortage is the actual cause of depression, or a result of it.

▷ Suffering from 'the blues'. Everyone has off days when they feel down or depressed.

△ Many fears arise from early experiences. This baby is being introduced to water in safety. A child who has an accident in water may become afraid of swimming.

The effects of exercise on mood are so positive that some people who work out a lot become hooked on their exercise routine. They start to feel guilty if they miss their exercise on a particular day. There is a danger that they might train too much, making their bodies more prone to illness or injury. Like most things in life, it is important to strike the correct balance between vigorous exercise and relaxation.

Our fears

People are often teased about being afraid of something. But fear can keep you alive. Fear makes you cautious about exposing yourself to danger. It is sensible to be afraid of sharks, poisonous snakes and suspicious-looking strangers. As a newborn baby, you are not born with many of these fears. But as you grow up, through stories, pictures and television, as well as through warnings from adults, you soon learn about the possible dangers that may be waiting for you around the next corner.

Why are we especially afraid of some things, such as a dark corridor or a creaking door? Why do we sometimes have nightmares about hairy spiders and monsters with fangs? In the modern world many more people are killed by cars and guns than by these creatures of the night – yet we do not have similar nightmares about cars. Perhaps as human beings have evolved, we have become wary of danger. We have developed fears about dangerous creatures and natural hazards. In the past, people who learned these fears quickly were clearly more likely to survive than those who were less cautious. The hunter who was constantly alert in case of a surprise attack by a wild animal was more likely to survive than the one who ignored such dangers. Perhaps modern hazards such as the motor car have not been around for long enough to influence our fears through evolution, in the same way as poisonous snakes have done.

Fear often has nothing to do with the actual risk involved in something. Being a passenger in a car is actually much more dangerous than flying in an aeroplane, yet many more people are afraid of flying than of travelling by road. Perhaps this is explained by our ancient fear of high places.

Phobias

Sometimes people develop an exceptional fear which is out of proportion to the risk from the thing they fear. It starts to affect their everyday life and the way in which they behave. We call this kind of fear a 'phobia'. Some people have a phobia about flying. Others have phobias about snakes, spiders, invisible germs, closed spaces or even being outside. Sometimes people with phobias overcome them by learning to relax their muscles and mind when they are exposed to their fear – sitting in an airplane waiting for take-off, for example.

Coping with stress

When something threatens to disturb our mental wellbeing and makes us worried or nervous, we say that we are 'under stress'. Stress can be caused by a sudden and unexpected event, for example a road accident, a robbery or a natural disaster such

---

DRUGS

Some people use drugs to alter their mood. Mood-altering drugs range from legal substances such as tobacco and alcohol, through solvents (chemicals in glue and other household products), to illegal drugs like cannabis, Ecstasy and heroin. Most of these drugs work in a similar way to exercise – they influence the release of neuro-transmitters in the brain. While drugs may produce a short-term pleasurable change in mood, unlike exercise they have no long-term health benefits. In fact, mood-altering drugs create many more health problems than they solve. Smoking tobacco causes lung cancer and heart disease, and alcohol can damage the liver. With all drugs there is the risk of becoming dependent on them. Some people become psychologically dependent on a drug, and are unable to cope with everyday life without the 'support' of their drug. Others become physically dependent, and their bodies do not function unless they can take the drug. In some cases the person may have painful withdrawal symptoms if they stop using the drug. There is also the risk of dying after taking a drug overdose or a contaminated illegal drug that has been mixed with some other dangerous substance. Most solvents are toxic (poisonous), and many children have died after sniffing them. Taking drugs affects the ability of your brain to react to danger or to make sensible decisions. For example, you cannot drive a car safely after drinking alcohol or taking any other drug that affects your perception of risk, or your judgement about your own abilities.

◁ It makes sense to be afraid of snakes – some of them are poisonous! People who learn that snakes are dangerous are more likely to survive than those who do not.

makes us feel alive and gives a sense of purpose to our lives. When the day of the event arrives, our minds focus on getting everything right so that we can do well. Our bodies respond by producing a hormone called 'adrenalin' (see page 58). It tunes up our heart and muscles, and gives us a nervous feeling of having 'butterflies in the tummy'. We must take care not to become too tense or we might 'freeze'. By breathing deeply and trying to make your mind go blank you can help yourself to remain calm. When the event starts, whether it is a race, a test or a performance, many people find that the stressful feelings of being nervous disappear as they get on with what they have been preparing to do.

Stressed out

Some people have jobs that expose them to traumas on a daily basis, for example fire-fighters and other emergency workers. The constant stress can produce health problems. People suffering from stress often have difficulty in sleeping, and their immune system is weakened so that they are more likely to catch infections. In the long term heart disease may result. Everyday causes of stress, such as noisy

neighbours and unhappy relationships at school or at home, can produce similar symptoms. We need to resolve these causes of stress if we are to feel well.

Taking care of your mind

You need to look after your body in order to stay healthy. You need to care for your mind too. When you have to face stressful events in your life, there are things you can do to help yourself cope with the stress: take exercise, make sure you have plenty of sleep, eat properly, organize your time, talk to friends. All these things will help you to keep a sensible approach to your problems. As you get older you can choose whether to keep your mind active with new projects and ideas, or let yourself become mentally lazy. Unhappiness and depression are often associated with a lack of purpose and with loneliness. People who make the most of their talents and take an interest in the outside world, who maintain family relationships and develop friendships, are more likely to have good mental health.

as a hurricane or an earthquake. Stressful events of this kind are called 'traumas'. People who experience stress following a traumatic event are often in shock. At first they are silent and dazed and may wander around aimlessly. Later they may keep reliving the traumatic experience in their mind. They are unable to concentrate on everyday things, they cannot sleep and they forget to eat or dress properly. Gradually, with the support of their family, friends or special counsellors, they talk about their experience and begin to accept that it has happened. By doing this, stress victims can usually start to live a normal life once again.

Be prepared!

We can never prepare for the shock of a trauma but our bodies and minds can prepare for the stressful events that we know about, such as entering an athletics competition or giving a public perfor-mance. A certain amount of this kind of predictable stress can be beneficial; it

▷ Some people deliberately expose themselves to stress in daring sports such as bungee jumping.

LIVING TOGETHER

We live in communities, from tiny rural villages to vibrant modern cities. Each of us must share our needs, desires and ambitions with billions of other individual human beings. Sometimes we compete, sometimes we cooperate. Every part of our lives is influenced by the ideas and discoveries communicated between individuals and groups, around the world and across the generations.

COMMUNICATION

From a smile and a wave, to a message on the Internet, communication keeps us in touch with each other. Our unique ability to use language puts us ahead of all other animals in the world of communications.

No living thing exists all on its own. Every organism shares its life with the rest of the living world. Communication is the transmission of signals from one living thing to another. It is part of the process of staying alive, of surviving. In the animal kingdom, most communications are connected with basic life processes such as finding a mate, warning off predators or keeping in touch with young offspring. The signs and signals that animals use to communicate include scents, sounds, gestures and displays. Some female moths, for example, release special chemicals called 'pheromones' which attract male moths several kilometres away. The arched back, snarl and open claws of an angry cat signal that no one should meddle with it.

Messages without words

Like other animals, human beings use non-verbal methods of communication in everyday dealings with each other. Sometimes we do this more than we realize. Your mother may not need to say anything to communicate her feelings when she reads your school report. The way she holds it, and the look on her face, immediately tell you if she is delighted, surprised, disappointed or angry!

Facial expressions and gestures are powerful ways of communicating our feelings and emotions. Some of these have the same meaning all over the world. For example, in every culture people's faces smile when they are happy or screw up tightly when they are disgusted. Showing your open palms is widely recognized as a signal of peace, while raising a clenched fist is a forceful signal of power or

aggression. Some gestures, however, have different meanings in different places. A friendly gesture in one culture can be seen as rude or obscene in another – so take care!

You can also signal your feelings by using sounds that are not words. Sighs and groans signal that you are bored, depressed or frustrated, and laughter tells others that you are amused or delighted; screams indicate that you are frightened, and wailing that you are full of grief.

Touch is another powerful way of communicating your feelings. Handshakes, kisses and hugs are different

▷ Body language, our gestures and the way we hold ourselves, can communicate almost as much as the words we speak. This man's open hands are showing that he wishes to be straightforward and honest.

△ Crossing your arms is a defensive gesture. It signals that you do not welcome the message that is being delivered.

△ In different circumstances a clenched fist can signal power, triumph, anger or aggression.

culture at one time. Perhaps it shows that the person is wealthy enough to eat well. Yet in another culture a slim body may be considered attractive and desirable, possibly because it is seen as a sign of fitness and youth.

The clothes we wear to cover our bodies communicate a lot about us. Smart, expensive suits are the choice of business people or politicians who want to appear serious. A white coat signals that a person is a doctor or a scientist; glamorous clothes are worn by people in the entertainment industry; while torn jeans and green hair are signs of rebellious youth. Your clothes send out a message about you. It is important to remember this when meeting someone for the first time, whether at a job interview or at a party.

Putting it into words

When you really want to make a message clear, or express a complicated idea, you need to put it into words. Human beings are the only animals who can produce a highly complex language. You may not fly like a bird or swim like a dolphin, but you are an expert communicator!

You may know tens of thousands of different words, but your spoken language actually consists of probably fewer than 50 separate sounds. We call these sounds 'phonemes'. The English language has about 40 phonemes, such as the 'a' in 'cat' or the 'p' in 'pin'. These phonemes have no meaning on their own, but we combine them to form different words. When you say the word 'cough', for example, you join together three separate sounds: 'c', 'o' and 'f'. In this way, with just a few sounds, you can make thousands of different words.

From words to ideas

Every word in our language has its own meaning. A word may stand for a single thing, or for a whole group or class of things in the world. The word 'house', for example, refers not just to one particular building, but to millions and millions of structures in which people live that we think of as houses.

When you talk you link words together into sentences to express ideas, or 'propositions', such as 'The sun is hot' or 'The

ways of greeting people. Hugs and kisses are also signals of affection, expressing love between friends, family members and sexual partners.

Fat or thin? Smart or scruffy?

Many animals have special body parts which they use to communicate visually to each other. The stag with his huge antlers or a peacock with his spectacular tail are just two examples. Our bodies are not specially adapted to communicate visually to the same degree as these animals. Yet male and female human bodies are different, and our secondary sexual characteristics (see page 72) are visual signals which attract the opposite sex. However, the same body shape is not always seen as attractive by every culture and every group of people. A large, well-developed body with bulky muscles or plenty of body fat may be the ideal shape in a particular

HOW MANY LANGUAGES?

There are thousands of languages in the world, but linguists (people who study language) suspect that all languages work in a similar way. All human beings seem to have a natural, or 'innate', ability to use language (see page 100). We all learn the language we hear around us as we grow up. One of the major differences between languages is their basic sounds. Young children seem to be especially sensitive to these sounds and can pick them up without difficulty. But after the age of nine or ten we find the sounds of a new language much harder to learn. The Japanese language does not have separate sounds for 'r' and 'l'. Japanese adults find it difficult to make these sounds when they learn to speak a language which does use them, such as English. English speakers learning Hindi experience similar difficulties with two separate 'p' sounds, which have no equivalent in the English language.

snow is melting.' You can link propositions to express more complicated ideas, for example: 'The snow is melting because the sun is hot.' By linking words to make sentences, you can communicate every possible idea.

From speaking to writing

Language has spread and changed as people have moved all around the world and settled in new areas. As language developed, people used it to tell stories about their ancestors, to pass on myths and legends about the past, and to imagine what life was like in distant places. Spoken, or 'oral', traditions were created in families and communities. These oral traditions often continue to this day, with parents and grandparents passing on favourite stories and family tales to their children and grandchildren. You probably know some family stories that are always told at family birthdays or weddings.

About 5000 years ago in Mesopotamia (see page 129) and ancient Egypt, people used writing for the first time. In the beginning they made symbols on clay tablets or parchment scrolls. Each symbol stood for a single object, such as a sheaf of corn or a jar of oil. The disadvantage with this system of writing is that you need a separate symbol for every word, so it soon became very complicated. Gradually signs for the different sounds that make up words were introduced. Chinese writing still uses a mixture of signs that stand for things, ideas and sounds. It developed about 4000 years ago and is probably the oldest writing system still in use. There are about 50,000 different Chinese characters, but most people get by with learning just a few thousand of them!

Alphabets

Most modern languages are written with an alphabet of letters with each one standing for a sound. Modern English has 26 letters based on the alphabet used by the ancient Romans. Unfortunately the letters of the Roman alphabet do not match exactly the forty phonemes we use in spoken English. For example, the 'c' in 'cat' sounds very different from the 'c' in 'nice', so learning to spell is not as straightforward as it might be!

With the invention of writing, people began to write down the things they knew, and so recorded history began. For most of history only a few people learned how to read and write – people such as royalty, priests, politicians and wealthy merchants. Today we believe that every child has the right to learn to read and write, and to gain access to the information and knowledge stored in books.

◁ The ancient Sumerians invented the first known writing system. It was based on wedge-shaped, or 'cuneiform', characters pressed into clay tablets with a stick.

The communications revolution

Until the 15th century, written documents were made and copied by hand. Then, in 1450, a great communications revolution began. In Mainz in Germany, Johann Gutenberg invented the first printing press with movable type (letters). This allowed books to be printed in large numbers. The first printed book was an edition of the Bible. With the mass production of books, they could be widely distributed for the first time.

Since the invention of printing, other revolutions have also taken place in communications. In the 19th century, a series of discoveries and inventions brought the far corners of the world closer together. People still sent written messages, but the invention of railways increased the speed of their delivery by mail. In 1837, the telegraph was invented. This sent messages along electric wires, at the speed of light. In 1858 the first telegraph cable was laid across the sea-bed of the Atlantic Ocean. In 1876, a Scottish inventor, Alexander Graham Bell, invented the telephone. In 1895, an Italian, Guglielmo Marconi, made the first long-distance radio transmission.

△ At a 'cyber cafe' you can use the Internet to exchange information and join in discussions with people from all around the world at the same time as enjoying a cup of coffee.

Space highways

Since then, technology has allowed us to launch communications satellites into space. Now with a small mobile phone that you can carry in your pocket you can talk to someone on the other side of the world. Perhaps the world's telephone network is the largest, most important machine that human beings have invented to date.

This network is now being combined with computer technology to give us access to information in every corner of the world. With a desktop computer which is plugged into a telephone line, you can keep in touch with people by sending and receiving electronic mail (e-mail) on the Internet (the global computer network). You can obtain information from databases that are stored on computers in other countries, and you can even present your own ideas to a worldwide audience from your school desk or bedroom. Communications technology is turning the world into a 'global village', in which we can all share in each others' lives and experiences.

◁ Before the invention of printing, important books were treated as great works of art. Only a few copies were made of each book and they were locked away in libraries.

FAMILIES AND GROUPS

Human beings are social animals. We spend our lives in various groups, from close-knit families to nations of millions. Cooperation between members ensures a group's survival, and we take comfort from them too. But living in groups can also lead to prejudice and conflict when one group feels superior to, or threatened by, another.

Some large animals, such as tigers and polar bears, live alone for most of their lives. The males and females must come together to mate, and the cubs are reared by their mother. But as soon as they can fend for themselves the mother drives her cubs away. Large solitary animals like these are comparatively rare. Most larger animals, including elephants, lions, wolves, whales, monkeys and human beings, live in social groups. This is a much more successful strategy.

There is safety and strength in numbers. The members of a group can cooperate to find food, share the burden of rearing their young and warn each other when

predators approach. A group can occupy a large territory, protecting its food supply from rival groups of the same species.

For most of our history, from our origins in Africa more than 100,000 years ago, we human beings lived in small bands, surviving by hunting and gathering. Then about 12,000 years ago, as we settled down, first to a way of life based on agriculture, and then developing industry and technology, the size and complexity of our social groupings increased dramatically. The arrival of agriculture meant that our ancestors no longer had to roam around looking for food. Growing crops and rearing animals allowed them to stay in

one place. Bands merged into tribes, people settled in large numbers in villages and towns, towns grew into cities, and, eventually, cities expanded to form nations.

Today, each of us may be members of several different groups. We probably 'belong' to a large human group that occupies a particular part of the world – we may be British, Spanish or Australian, for example. Our national identity often determines the language we speak, the foods we prefer to eat and the clothes we wear. But equally important are the many smaller social groups we belong to, which give structure and meaning to our lives. These groups include our friends, the people who live in the same neighbourhood, people who go to the same school, share the same hobbies, attend the same church or belong to the same sports club. Above all, there is our family.

Family

Have you ever drawn your family tree? Family links are often important in giving us a sense of who we are and where we come from. Each one of us has a biological family with whom we share genes (see page 124): two parents, four grandparents and often brothers, sisters, aunts, uncles, cousins and other relatives. As adults we may have children of our own. Within a family biological ties are very strong, but there may be conflict between any of the members, especially as children mature and seek to establish their own personalities. Sometimes family relationships do break down.

Patterns of family life vary tremendously. Some children are brought up by both parents, some by one biological parent, by foster parents, or by grandparents. In some affluent societies children over seven may be sent to a boarding-school for much of the year, returning to the family home during the

◁ The popular image of the nuclear family portrayed by magazines and television is only one of many possible family structures. It is less common than the media might suggest.

holidays. In modern Israel the children on a kibbutz (a kind of communal settlement) sometimes live in special children's houses, not in their parent's house. Children who, for one reason or another, are separated permanently from their family may grow up with a deep curiosity about their biological parents. Even if they are happy and secure with loving foster parents, they may have a strong desire to discover their biological origins.

The structure of the family has varied greatly though history, and in different societies around the world. The 'nuclear' family consists of parents and their children. It is a common family group in industrial societies today. In contrast, in an 'extended' family grandparents, children and grandchildren all live together. The head of the family is usually the senior male, more rarely the senior female. Sons stay in the family home when they marry and have children. Daughters usually leave home to marry into other families.

On the death of the senior male, the eldest son becomes the new family head. This was once the common pattern in many parts of the world, including Europe, when family life was based around the ownership and farming of a particular plot of land.

The European royal families are good examples of extended families, headed by a senior male or female, with strong links maintained through several generations. Today, extended families are still common in rural communities in developing countries such as India, but in industrial societies, where people must move frequently to find work, extended family ties are more difficult to maintain.

In most western countries it is usually illegal for a man or a woman to marry more than one partner at a time. This arrangement is called 'monogamy'. Sometimes the relationship between two partners breaks down, in which case they may divorce and marry new partners. The

pattern in which an individual takes several partners in succession during his or her lifetime is called 'serial monogamy'.

Monogamy is not the only possible arrangement for marriage however. Some societies allow a man to have more than one wife, or a woman more than one husband, at the same time. 'Polygamy' is the name given to the custom whereby a man marries several wives. It is practised by many African tribes and was once common among the Mormon religious community of North America. It is also permitted by Islamic law, which allows a man to take up to four wives.

'Polyandry', when a woman marries more than one husband, is less common than polygamy. Women of the Nair people of India may marry several husbands at the same time, and family property is inherited through the female line. In some communities in Tibet it is the custom that when a woman marries a man she also takes his brothers as husbands.

SELFISH GENES

Why do we feel a close bond to family members, often despite personality clashes and conflicts? One important biological reason is that we share many genes in common. It is in our interest that we cooperate with and take care of each other, so that those genes may stand a better chance of surviving than the genes in solitary animals. This 'selfish gene' theory goes a long way to explain why parents take care of their children, and perhaps why we often care more for our close relatives than for other individuals.

You inherited half of your genes from your mother and half from your father. According to the selfish gene theory, they make sacrifices to take care of you in order to ensure the survival of their genes. We are certain that half the genes in our bodies are identical to those of each of our parents, and to those of our children. We also share a proportion of our genes with other members of our family. Selfish genes lead us to be more concerned for the welfare of close family members than more distant ones.

We share half our genes with each of our parents, and on average, we share half with our brothers and sisters, a quarter with our grandparents, uncles and aunts, and an eighth with cousins. These genetic links reflect the closeness of the concern we have for our relatives.

Bands and tribes

In hunter-gatherer societies several families banded together to form groups of perhaps 25 individuals. The size of the group allowed its members to share important tasks, such as food gathering, and provided strength and security. Although not all one family, the members of a band would have been quite closely related, perhaps able to trace their origins to a relatively recent common ancestor. The San people of South Africa still follow this lifestyle today, but most of us now live in much larger communities. Our hunter-gatherer past, however, leaves us with an important legacy – we feel most secure

living and working in groups of, at most, a few hundred individuals we know and recognize. Schools seem to work best with about 600 pupils. Directors of large companies with thousands of employees have realized that people work most effectively when they are part of smaller groups with well-defined responsibilities and goals.

About 12,000 years ago, when wandering bands of hunter and gatherers began to settle and farm, alliances would have formed between neighbouring bands. An area became the home of a tribe or clan, perhaps consisting of several thousand individuals. As social groups became larger more organization was required. An

overall chief or a council of elders would be appointed, to organize the collection and distribution of food, defence of the tribal home, and perhaps to plan raids against neighbouring tribes. Membership of clans or tribes can often be traced through a common name passed on through the generations. For example, people throughout the world with names such as Campbell or Macdonald can trace their origins to Scottish clans.

Tribal bonds are still strong in many parts of the world today. The Zulu are one of the best-known tribes of South Africa. About 2 million Zulu live in the province of Natal. Their way of life is based on growing millet and raising cattle. Their

traditional homes are villages of beehive-shaped huts, grouped in a circle inside which the cattle graze. Many Zulu have now moved to find work in mines and cities. A Zulu-based political movement called Inkatha has played an important part in recent South African politics. It has come into violent conflict with the African National Congress, which won the first free elections in South Africa in 1994.

Sticking together

Anthropologists study human groups to try to discover how they are organized and what keeps them together. They look at the customs, beliefs, language, relationships and codes of behaviour of a particular group. As well as studying tribal peoples, they work with groups in companies, religious organizations, schools and colleges.

Lifestyles vary enormously, but all human groups have elements in common. For example, an invitation to share a family meal is universally acknowledged as an offer of trust and friendship. Shared beliefs and knowledge passed on through stories and rituals, common interests or a common goal, keep a group together. Expected patterns of behaviour must be followed, and individuals who break the rules may be punished, perhaps expelled from the tribe, or even killed.

Some members of the group are leaders, others are followers. There may be a single chief from a particular family, or a council of wise elders who take decisions. Historically, success in battle was often the requirement for leadership. Warrior chiefs such as the Mongol warlord Ghengis Khan, whose armies conquered China in the 13th century, led through strength, fear and cunning. In the modern world knowledge, wealth and the ability to influence the ideas of others, are more common leadership characteristics.

RITES AND RITUALS

Ceremonies and rituals mark important events in the life of a group, reinforcing the sense of common identity. 'Rites of passage' are rituals associated with important stages in the lives of individual group members, such as birth, marriage and death. Many groups have rites of passage linked with puberty. These mark the entry of the young person into the world of adulthood. For example, at puberty the young men and women in some native American tribes were sent to fast and pray alone in the wilderness. Eighteenth birthday parties, graduation from school or college, or starting your first job could all be considered as modern rites of passage.

'Initiation rites' often mark the entry of an individual into membership of a particular group. After a period spent learning the 'secrets' of a craft or profession, new members may have to swear an oath to uphold the group's high standards.

Feast days and festivals mark events that affect a whole community, such as the gathering of the harvest, the longest day of the year or an important historical or religious event. At the great Mardi Gras carnivals in Latin America people enjoy themselves before the more solemn periods of Lent and Easter. In the United States, Independence Day on 4 July celebrates the declaration, in 1776, of the independence of the former British colonies. Sometimes the original reason for a holiday may be just a vague memory. For example, the origins of May Day festivals were probably ancient ceremonies in honour of Flora, the Roman goddess of spring. Today in northern Europe, May Day is celebrated as a festival of new life when children wear flower garlands and dance around the maypole.

△ A Masai boy in Kenya has his hair cut by his mother as part of the initiation ceremony that will take him into manhood.

◁ The group of friends we make at school may remain our friends throughout our lives. Shared memories of growing up are a powerful bond that can be recalled and enjoyed when we meet again in adult life.

group will be approved of by another. When we visit another country we should make ourselves familiar with local customs if we are to avoid, even intentionally, causing offence. In the West, for example, eating noisily is considered bad manners, but in the East it shows that you are enjoying your food and is therefore taken as a compliment to the person who provided it.

Our behaviour in a group can be very different from our behaviour as individuals, perhaps because in a group we feel less of a sense of personal responsibility or risk. The consequences can be both positive and negative. Many people enjoy singing in public as members of a choir, but would not dream of standing on the stage alone to perform. Similarly, it is much easier to rise to the challenge of demanding activities such as mountain-walking, canoeing or parachute-jumping with the support and encouragement of others. On the other hand, the friendly 'tribal' conflicts between rival supporters at football matches sometimes get out of hand and become violent. Many of the

Join the club

Membership of any group has its benefits, but it also has its responsibilities and limitations. You cannot behave just as you wish and still hope to be accepted. If a group member breaks the rules, then at best they will become unpopular, at worst they may be asked to leave or be expelled. In sport strict rules are essential if a game is not to disintegrate into chaos. Football players may be sent off the pitch in the middle of a match, and even suspended from playing for the rest of the season, if they persist in behaving badly on the field.

Most groups have agreed rules or laws on how to behave. But it is a mistake to assume that what is acceptable in one

▷ A church congregation in Sydney, Australia. Religious groups share beliefs about the nature and meaning of their lives. Regular acts of worship bring the group together.

people who take part in this violence would not dream of behaving in this way if they were not influenced by the actions of the group as a whole.

Conflict

In the natural world creatures compete for territory and food. Predators must kill their prey. Individual members of a species must compete for mates. Males display their strength and, in many species, fight to establish a territory and attract females. As part of the natural world, humans have also evolved powerful competitive and aggressive instincts. Within most human groups competition is strong, as each individual tries to assert their place and gain the favour of others. The advantages of co-operation between members of a group and the fear of reprisal usually keep individual violence in check. But the instinct to fight, particularly between young males, can sometimes emerge. In modern society this is frowned upon and individual

aggressive instincts are channelled into sport, work and other non-threatening activities.

We might have hoped that the human being's capacity to reason, and to choose to act together for the common good, would allow us to rise above our 'animal' natures. But competition between human groups for food, land, resources and power still leads to conflict and aggression. The 20th century has seen some of the most terrible wars in history, with more sophisticated and destructive weapons being used than ever before. Since the atomic bomb was dropped on Hiroshima in Japan, on 6 August 1945, fear of a global nuclear war has perhaps helped to prevent war between the major industrial nations. But the world is not at peace. At any one time in the 1990s there have been as many as 40 different wars raging around the world, both between nations and within them (civil wars). Some conflicts are driven by economics. For example, an important factor in the Gulf War of 1991 was the western world's dependence on the oil

supply from Kuwait, which had been cut off by Iraq's invasion of that country. Other wars, such as the conflicts between the peoples of the former Yugoslavia, are driven by fear, racial hatred and memories of past wrongs passed on through the generations.

How can conflict be avoided? Perhaps the greatest enemy of peace is ignorance. Ignorance of other people and their culture, breeds suspicion and fear. Recent studies of human DNA (see page 12) have revealed that human beings throughout the world are incredibly similar – we have evolved from common ancestors and share common needs and problems. Perhaps this knowledge will help dispel some of the myths about differences between human groups. The information revolution, which has now linked all parts of the globe through satellite communications, and the 'information superhighway', should also help us to understand each other better. But we still have a long way to go if we are to create a peaceful world.

◁ Demonstrators need not resort to violence to express strong feelings about an important issue.

From Settlements to Cities

Just 500 to 600 generations have passed by since our ancestors first learned to farm. During this time, human settlements have grown from simple villages to 'high-tech' cities, where millions of people live and work, and where money, food and goods are traded across the world.

The lifestyle of a farmer is harder than that of a hunter. A farmer devotes many more hours to growing crops on the land and to tending animals than a hunter spends tracking wild antelope or bison for food. The comparison between the hunter and farmer on the one hand, and a lion and a bee on the other, is a similar one. Lions spend most of their time sleeping in the sun, but bees are working all the time to build their hives and collect nectar to make honey. So why would a hunter-gatherer give up that lifestyle to settle on a farm? The answer is that farming can support many more people. It is probably a safer way of life, too.

Life on the farm

The first farmers we know of settled in the Middle East and North Africa about 10,000 years ago. The remains of their villages have been found in the region known as Mesopotamia around the Tigris and Euphrates rivers, along the shores of the Mediterranean Sea and on the banks of the River Nile. These areas are now parts of modern Iraq, Israel and Egypt. One of the earliest settlements recorded by history was at Jericho where, around 9000 years ago, perhaps 2000 people lived. There has been a town or village here ever since.

Agriculture, along with tool-making and language, is one of the greatest discoveries made by humans. To be a successful farmer you must have a clear plan for the future, save seeds, plant them and tend the crops as they grow. You do not reap the rewards of your hard work until many months have passed, when you finally gather in the harvest.

Sowing the seeds

The first crops were probably grown by accident. People noticed that split grain grew where it was dropped and had the idea of sowing it closer to home.

The first farmers gathered and sowed the wild grass seeds that grew naturally in their area. Today, by far the most important foods we eat are the cereals that farmers have developed from wild grasses. This has been done by many generations of selective breeding (see page 12). Cereals such as wheat, rice, corn, rye, oats, barley, sorghum and millet feed most of the world's population.

As farmers settled on the land they learned to domesticate animals. The bones of small cattle have been found at village sites which date from about 8000 years ago. Sheep, goats and pigs were kept for their meat and skins. But it was not for several thousand years that humans discovered how to make animals work for them in the fields. The first pictures of animals at work carrying loads or treading seeds into the soil come from Egypt about 5000 years ago.

◁ Subsistence farming is back-breaking work. All your time and energy are devoted to growing the food you need to survive. These farmers are planting rice, one of the world's most important subsistence crops, in flooded paddy-fields.

Food to spare

The first farmers must have added to their food supply by continuing to hunt animals and by gathering wild plants and fruits. Eventually, as their farming skills improved, they could grow enough food to supply all their needs. With more food people began to live longer – food could be stored for hard times in the cold dark winters or long dry summers, and people who could no longer find their own food, such as the sick and elderly, could be fed. Living off the food that you grow yourself is called 'subsistence farming'. Nearly half the world's population still survives in this way today, living off the food produced in their local rural community.

If the soil and climate in a particular area are favourable, the local farming community can usually produce more food than it needs. This extra food is called a 'surplus'. In places where there was surplus food, some people started to move away from working on the land. They began to specialize in activities such as tool-making, pottery manufacture or weaving. The people who worked then began to divide up into different groups. With this division of labour, some people became full-time organizers or leaders of the community, others were warriors who defended it, or merchants who traded the surplus crops and manufactured goods, or wise men, politicians, soothsayers (people who predict the future) and priests. The community needed laws to help it run smoothly. People then found that they had some spare time to spend inventing ways of recording events, to make music and to experiment with new materials such as metals and glass. These activities were the beginnings of what we call 'civilization', and they took place as villages grew into towns and towns expanded into cities.

THE FIRST CITIES

The first civilizations grew up in Mesopotamia about 5500 years ago. People called the Sumerians built walled cities of houses, shops and palaces from clay, around pyramid-shaped temples called 'ziggurats'. The Sumerian civilization lasted for many centuries, and its people created fine works of art including carved stone and metal statues. We know quite a lot about life in the first cities because the Sumerians recorded events on clay tablets with their cuneiform (wedge-shaped) writing (see page 120). Each city had its own king. The temple held an image of the city's god to which the people made offerings. One 4000-year-old tablet records that children went to school and were beaten with a cane if they were late, talked or did not do their lessons properly!

In the following 4000 years civilizations grew up all around the world as human beings independently discovered agriculture and became more wealthy. Civilizations arose in Egypt, India, China, Greece, Rome and Central America. As the populations in these places increased, so did the power and influence of leaders. Armies were formed to conquer neighbours and the influence of certain cities expanded to create whole empires. Within individual civilizations rival groups with different philosophies or beliefs plotted and fought against each other; kings, emperors and different forms of government came to power and then fell. The struggle for political power which still continues to this day had begun.

▽ The ruined Inca city of Machu Picchu lies high in the Andes Mountains in Peru. It was discovered in 1911. Little is known of its history, but like the cities of much older civilizations in the Middle East and Asia, many of its buildings seem to have had a religious significance.

Making a living

During the 20th century the division of labour has developed so much that each person now does very few of the basic tasks that are needed to keep us alive. You may spend your whole life without ever hunting or growing your own food. You may never have to build your own home, make your own tools, collect your own fuel, find your own water supply or make your own clothes. Our hunter-gatherer ancestors had to do all these jobs for themselves. As our communities have grown, so our individual roles have become more and more specialized. We are no longer just human beings; instead we call ourselves farmers, teachers, engineers, doctors, lawyers, soldiers, shopkeepers, artists, students, scientists and so on.

Early farmers used to trade their surplus goods at village markets. They used a method of exchange called 'bartering', where goods of the same value are swapped without anyone paying money. Today, most people make a living by working for money, which we then exchange for the things we need. Our complicated society is kept going by this trade.

What you own

Trade depends on the idea of ownership. We feel that we own the things that we grow or make – they belong to us. If you spend hours modelling a lump of clay into a pot, or carving a figure from a piece of wood, the object becomes your own. Ownership is an extension of the territorial instinct that so many animals possess. A pack of wolves, for example, defends a certain area of land as its territory. The wolves keep out rival packs by howling and marking the boundaries of this territory with their scent. The territory is important because it provides the pack with the food it needs throughout the year.

The idea of trading the things that we own is a human invention, however. Our intelligence allows us to work out the value of different things by considering the effort needed to find or manufacture them, and their future value to us. For example, most people would agree that one hour's worth of chopping wood could be worth a square meal, but two eggs are not worth a horse! Yet people from different cultures do not always agree on the value of something. In 1626 a Dutchman called Peter Minuit bought the island of Manhattan (New York) from the Native Americans who lived there. He paid them with beads and trinkets worth just $24! Today Manhattan is one of the most valuable pieces of land on Earth.

Money, money, money

Modern trade takes place across the world. In a British supermarket you can buy apples from New Zealand, bananas from the Caribbean, tea from China and coffee from Brazil. The method of exchange that makes this worldwide trade possible is not bartering but money. Money is a very sophisticated idea which the human mind invented. Just as we invented words to represent things and ideas, so we have invented money to represent value. Money is coins, notes or other tokens that you can swap for goods or use to pay other people to do things. The first tokens used as money were probably small items such as shells and beads. The oldest known metal coins come from Turkey and the first banknotes were made in China nearly 3000 years ago.

City life

A modern city is a very complex system. There are now more than 250 cities in the world with a population of 1 million or more. Each year more and more people move from the countryside into cities to find work. To keep an enormous city of more than 10 million people working, such as London or New York, needs an incredible amount of organization. Every day more than 1000 lorry loads of food must be brought into the city just to provide that day's meals. Thousands more lorries, as well as numerous ships, aeroplanes and trains, are needed to bring in goods to be traded and raw materials to be processed in the factories.

Thousands of millions of litres of fresh water must be pumped into the modern city every day. Rivers of sewage have to be treated and made safe before being put back into the environment. Thousands of tons of rubbish must be collected and

▽ We make a living from trading our wares in markets. Today the word 'market' describes many kinds of activity – from traditional street markets like this one, to the job market where we sell our talents and skills, to international stock markets where shares in huge companies are bought and sold.

disposed of safely. Electricity from huge power stations has to be transmitted along tens of thousands of kilometres of underground cables to light and heat shops, offices, hospitals and homes. Road and rail transport systems must be organized to move people around.

Just another day

Each day in a modern city like New York or London, 2 to 3 million children and young adults go to school or college to be taught by tens of thousands of teachers. Hundreds of new babies are born and

hundreds of people die; thousands more people have accidents or are taken ill and go to hospital. Every day thousands of crimes are committed, hundreds of criminals are arrested and then taken to court where they are fined or sent to prison. Millions of pounds' worth of goods are bought and sold in the city's shops, and hundreds of banks keep track of what happens to all the money. At night the activity continues. People go to theatres, sports events, cinemas, pubs and clubs.

City life would not be possible without our ability to plan, organize and communicate. A human city is a far more complex place than even the most highly developed colony of social insects like ants or bees. It is a place where millions of individual living animals – human beings – survive by finding ways of living together.

◁ Modern city dwellers like these people in Tokyo, Japan, are part of an incredible system which must provide food, water, energy, transport, health care, education, entertainment and many other resources on a vast scale.

▷ Cities attract people from poor rural communities because they seem to promise a more prosperous life. But there is not always work or a welcome for everyone. The contrast between the wealth of the city and the poverty of the shanty towns around it is dramatic.

RELIGION AND BELIEF

Why was the world created? What is the purpose of our lives? For many people, religion provides answers to these ultimate questions.

Human minds probably became aware of their own existence at some stage during prehistoric times. Ever since, people have tried to explain where we come from. Why are we alive, and what happens to us when we die? Throughout history, religion has provided some answers to these questions, usually through belief in an individual god or a group of gods. There are many different views about gods, but most religions worship one God, who is either an all-powerful being or a non-physical spirit, or force. Followers of these religions believe that God was responsible for creating our physical world, and can influence events in our lives. They believe that we can have a personal experience of God and that God helps us to know the difference between good and evil, and right and wrong.

Which faith?

History shows that religious beliefs have been held by all kinds of people, from bands of hunter-gatherers to great civilizations such as ancient Rome. The religious ideas of a people are often taught by priests or holy men. People meet to share their religion, perhaps by a special rock or in a special building, often on a particular day of the week. They worship their God in special ceremonies, which may include ritual (religious) music, dancing and sometimes sacrifices. Religious followers may study holy documents or books that contain the teachings of great religious thinkers of the past. Prayers and meditation help worshippers to think carefully about their beliefs, and to form a close and personal relationship with their God.

Tribal religions, such of those of the Native American tribes or the Aboriginal people of Australia, are followed by certain groups of people living in a particular place. Often they are concerned with the spirits, perhaps of dead ancestors, which are thought to be present in the animals, trees, rocks and rivers of the local environment.

Some religions have spread to people of many different countries. These are the great world religions. Today they include Buddhism, Christianity, Hinduism, Islam, Judaism and Sikhism. About 33 per cent of the world's population are classified as Christians. Islam is the next most widespread religion, with a following of 18 per cent of the world's population.

Common beliefs

Every religion has its own set of beliefs and explanations, but all religions share some common ideas. For example, most religions have a story of how human beings first came to exist. The ancient Greeks believed that the goddess Athene breathed life into tiny clay models of the gods made by Prometheus. The first book of the Jewish and Christian Bible tells how God created the world and all living things in six days. Stories such as these are known as 'creation myths'.

Another common religious idea is the belief in a spiritual world, where God lives and where our 'souls' or 'spirits' may travel when we die. Christians believe that our souls live on after we die and, if we have lived life in the right way, go to join God in Heaven. Hindus believe that after death our spirits are born again into another body. This is called 'reincarnation'.

Holy books and holy people

The teachings of the great religions are passed on through holy books or writings called 'scriptures'. The Koran is the holy book of the Islamic faith. Muslims (followers of Islam) believe that the Koran contains the words of Allah, their God, which were spoken by the angel Gabriel to

◁ Muslims pray five times each day facing towards Mecca, the birthplace of the prophet Muhammad.

▷ Pages from an ancient copy of the Koran, the holy book of the Islamic faith.

the prophet Muhammad as he prayed. The 114 chapters – called the Suras – of the Koran contain rules which Muslims must follow in their daily lives. These rules include instructions about prayer, hygiene and the preparation of food, as well as laws on marriage, inheritance and justice. Muslim children learn many of the Suras of the Koran by heart.

The origins of other great religions can also be traced to a single teacher or prophet such as Muhammad. The Christian faith is based on the New Testament of the Bible, which tells the story of the life and teachings of Jesus of Nazareth. He lived in Judaea (modern Israel) 2000 years ago. Christians believe that Jesus was the son of God who sacrificed his own life to take away evil from the world.

Judaism

Jewish beliefs are based on the story of the Jewish people which is told in the Hebrew Bible. Christians call this story the 'Old Testament'. The same stories also form the historical background to the Islamic faith. Jewish people believe that there is one God and that they have a special relationship with him. This relationship, or 'covenant', was made between God and Abraham, considered the father of the Jewish people. The first five books of the Bible tell stories about the creation, Adam and Eve (the first man and woman), and Abraham, and set out rules which Jewish people should follow in everyday life.

Hinduism

Hinduism, like Judaism, is an ancient religion that began in early civilizations more than 4000 years ago. It is the main religion in India. Hinduism calls for a very varied and tolerant way of life. Hindus believe that there are many ways to reach Brahman (God). These include good deeds, prayers and meditation. They believe that every living thing has a soul, which is a part of God, and that we should not wish to cause it harm. When living things die, their souls are reincarnated. The form that a person's soul takes when it is reborn depends on their deeds, or 'karma', during their lifetime. The Hindu God appears in many different forms, and is worshipped as a male, a female, a half-human and half-animal being, or even as fire.

Sikhs

The Sikh religion was founded by the guru (teacher) Nanak from northern India. Nanak believed that God was more important than religious differences. He taught that Sikhs should learn about God and remember him frequently with prayers. They should work hard and honestly and share their earnings with others, regardless of their religion or race. All Sikh men and women are equal, and there are no special priests or monks. Today, you can recognize Sikh men by the five religious symbols they wear: uncut hair (often

wrapped in a turban), a steel bracelet, a hair comb, white shorts (usually worn under other garments) and a sword (to defend the truth and help the weak). Today the sword may be a tiny symbolic dagger worn in the hair comb.

Buddhism

Buddhists follow the teachings of Prince Siddhartha Gautama (Buddha), who lived in India 2500 years ago. The word 'Buddha' means the 'Enlightened One'.

△ Buddhists from Tibet spin prayer wheels containing written prayers. They believe that the prayers take effect as the wheels turn.

Buddha taught that people could find peace and happiness, called 'Nirvana', not by worshipping gods, but by living life in the right way. As all living things are connected, if we harm another living thing we harm ourselves too. Buddhists look for the truth by meditating and by acts of kindness and compassion.

Sects and cults

Sometimes there are disagreements between the followers of a religion. One group, for example, may believe that only men should be priests, while another believes that women should be priests as well. These disagreements may cause some followers to form a new group, or 'sect', which practises the same religion but in a different way. The Christian religion, for example, has divided into many branches including the Roman Catholic Church, the Orthodox Church, the Anglican Church and the Methodist Church.

Cults are much smaller religious groups, with particular beliefs that are often very different from the teachings of the widespread 'established' religions. Cults are often influenced by the teachings of a particularly powerful or 'charismatic' leader, who claims to have special new knowledge of God's plan. Some cults are secretive, and the members cut themselves off from the rest of society, including their friends and family. In some cults, members may believe that the world is about to end or that the rest of society is evil. Recently there have been some tragic events involving cults: under the influence of very persuasive leaders, cult members have committed mass suicide or come into conflict with government authorities.

Science and religion

Modern science has provided some alternative answers to the very searching questions about the creation of the world and the beginning of life. We now know that the Universe began with the 'Big Bang' 15 billion years ago. The planet Earth formed from the dust of exploding stars about 5 billion years ago. Human beings, along with all other living things on Earth, were not created exactly as they look today. Living things have evolved gradually over 3.5 billion years (see page 11). Scientists have still not produced any evidence about a separate spiritual world, the possibility of life after death or the existence of God. Today many people choose to be 'agnostics', which means they accept that they do not know whether God exists. Others are 'atheists' who believe that there is no such thing as God, and that everything is part of the physical world which science can observe and measure.

▽ A religious procession in Spain. Holidays and festivals which mark religious events give structure to the year. Even if you are not especially religious yourself you can still look forward to sharing some of the traditions, such as giving presents and street parades, associated with special religious days.

Science has provided us with many new explanations about the world and our existence in it. Yet despite such great successes, many people, including some practising scientists, believe that there is more to existence than atoms and molecules. Some people's beliefs are based not on traditional religious teachings or scientific facts but on some kind of personal religious experience. They feel something wonderful at work in Nature which they call God.

Many people choose to express and share with others their deep feelings of respect and wonder in the Universe by following one of the great religions. In contrast, others look at the evil things that have happened in the name of religion and reject all religious faith. Some people have a sceptical attitude, saying simply 'I just don't know'. You may be lucky enough to live in a society where you can choose what to believe and which religion to follow without fear of being persecuted. Many other human beings across the world, both in the past and also still today, have not been so lucky.

CONFLICT AND PERSECUTION

The great religions can be powerful forces for good and for peace. Sadly, throughout history, they have also been associated with corruption, persecution and conflict. Religious leaders have sometimes used their position to become wealthy or gain political power. In the past, Catholic Christians have fought with Protestant Christians, and Muslims have fought with Hindus. Religious differences are still an important factor in modern conflicts, such as those in Northern Ireland and Bosnia-Herzegovina. The Jewish people have been persecuted through the ages since they were exiled from their homeland by the Romans in AD 135. Their persecution reached a tragic peak before and during the Second World War. In the 'Holocaust' of Nazi Germany more than 6 million Jews were put to death in concentration camps. In 1947 the United Nations established the Jewish state of Israel and millions of Jews have now returned to the original homeland. This has not been a peaceful process, however. The new Israelis displaced many Palestinian Arabs from what had become their homeland too. This has led to many years of conflict between Jews and Arabs in the Middle East.

△ The symbol of Amnesty International, a voluntary organization that investigates abuses of human rights, and campaigns on behalf of victims of political and religious persecution throughout the world.

▷ Orthodox Jews pray at the Wailing Wall in Jerusalem. Jerusalem, the capital of the modern Jewish state of Israel, has deep spiritual significance for Jews, Christians and Muslims. It has a long history of religious conflict.

CULTURE

The food we eat, the music we listen to, the clothes we wear, the books we read – these are just some of the things that make up our culture. Sharing a culture helps us to feel that we belong to a particular group of people.

If you are brought up in a western culture, you will probably eat your food with a knife and fork from a plate, greet friends and family by shaking hands or kissing, and give presents at Christmas. If you are brought up in an eastern culture, you might eat your food with chopsticks from a bowl, greet people by putting your palms together and bowing, and give sweets and flowers as presents to celebrate the New Year.

The differences between the cultures of human groups that have lived separately for many centuries are fascinating. By reading books and watching films and television we know more than ever before about cultures from different parts of the world. We experience different cultures when travelling abroad on holiday. We eat food from different cultures, buy clothes influenced by different cultures and listen to music from around the world. Are you a reggae fan, for example, or do you prefer a Latin American beat? We even buy objects from different cultures to decorate our homes.

Cultural wealth

As people have migrated around the world in search of work or a new lifestyle, they have taken their own culture with them. Today, in many of the world's major cities you can find a rich variety of cultures. This multicultural mix is sometimes called a cultural 'melting pot' and it is producing some fascinating new developments. In music, for example, young people are using modern technology to mix sounds from different cultures. Native American chants are mixed with modern dance music, Latin American rhythms with the folk melodies of Ireland, traditional Indian music with jazz. Now we are learning to appreciate and enjoy the differences between various cultures, rather than thinking of them as strange or even threatening.

▽ The groups of dancers and drummers at the annual Carnival in Rio de Janeiro, Brazil, spend many months preparing their fabulous costumes and exciting routines. Sharing in the traditions and great events of a culture makes individual human beings feel that they belong to something important.

Young and old

Have you heard of the expression 'youth culture'? Young people develop their own styles of dress, have their own music, magazines, TV programmes, dance styles, clubs and heroes and heroines. Their parents may not understand these tastes. Have your parents ever complained that your music is just noise, or your clothes are scruffy? Young people are united by their shared appreciation of style. If you are not up on the latest records or if your clothes are not 'cool', you probably feel left out.

Today's youth culture may seem quite different from your parents' style when they were young, but it is probably not as different as you think. If you look back in time you can see how today's music and dance styles, kinds of haircut and ideas have evolved from those of previous years. Some styles even come around again; for example, the music and clothes of the 1960s and 1970s are popular again in the 1990s. Youth culture is really just a fast-moving part of a larger and more permanent culture which is shared by all generations in the community. Most of us are born into a culture with long traditions which have been passed down from one generation to the next.

Art and beauty

One of the main ways in which we express our culture is through the arts. Art seems to have started sometime after modern human beings began to spread out of Africa (see page 22). The first known art forms were cave paintings and small figures and animals carved from ivory, bone and stone. Music and dance almost certainly date from ancient times too. Bone whistles and flutes have been found in prehistoric sites up to 25,000 years old in Hungary and Moldova, and every recorded modern human culture has some form of music-making.

Today art can be divided into several different branches. Paintings, sculpture, music, poetry, theatre and dance, as well as other art forms, including some architecture, which are produced for their beauty alone, make up the 'fine arts'. The 'decorative arts' are concerned with creating beautiful objects which also have a use, for example pottery, fabrics and furniture.

△ The pop singer formerly known as Prince. Modern popular music has helped to create an international youth culture with stars and styles known throughout the world.

Objects such as these are often decorated with traditional patterns that can be traced back a long way in history.

Literature

One of the most important ways in which culture is transmitted is through its literature. Literature consists of the poems, plays and books which express ideas, and tell stories, reflecting the experiences and beliefs of people living in a particular time and place. Many ancient civilizations had mythical tales of gods and heroes often performed or written down as long poems called 'epics'. The Greek writer Homer, for example, tells the story of the wars of Troy and the deeds of Achilles and Odysseus in his famous epics the *Iliad* and the *Odyssey*. In the Middle Ages popular tales called 'romances' recounted legends such as that of King Arthur and the Knights of the Round Table. The first great book that we

CULTURE CLUBS

People who migrate from one culture to another sometimes like to maintain links with their original culture – their roots. They do this by forming societies or clubs where they can meet. All over the world there are people who can trace their origins to Scotland, for example. In their new homeland the members of the Scottish society meet to try to keep alive some of the traditions of the culture they have left behind. From New York to Moscow there are meetings of Scottish groups at which the men dress in kilts, and people play on the bagpipes, recite the poems of Robert Burns, eat haggis, drink whisky and perform traditional Scottish dances.

Some people become so fascinated by a culture from the past that they try to re-create it in their homes or with friends. Some people in Europe are particularly fascinated by the way of life of western cowboys, for example. This may be because so many Hollywood films and TV series have glamorized the lifestyle of a cowboy, which in reality was probably hard and unpleasant. People who become obsessed with the popular image of the 'Wild West' spend their weekends dressing as cowboys, eating cowboy food and singing cowboy songs.

would recognize as a novel is probably *Don Quixote* by Spanish writer Miguel de Cervantes (1547–1616). It tells the story of a man who believes that romantic tales of knights and dragons are true, dresses in armour made from pots and pans and rides into battle against a windmill! Today tens of thousands of novels telling every kind of story from romance and crime to horror and science fiction are published each year.

A question of taste

Human ideas of beauty are not always the same. You have probably learned some of your ideas about beauty by growing up in a particular culture. Look at dance, for

◁ We can all appreciate the dramatic make-up and the beautiful costumes of traditional Japanese theatre, but it requires some study to understand the subtle meanings of the players' expressions and actions.

◁ The tradition of dance in Bali, which has lasted for 1000 years, is closely linked to the beliefs and spiritual life of the community. Music and dance are important parts of religious worship in many cultures.

example. If you have never been to the ballet you may not appreciate its beauty when you first see a performance; it might even appear ridiculous. In the Hindu religious dance Bharata Natyma, every movement has a meaning, but you cannot understand this meaning if you have never watched this kind of dance before. If you take the trouble to study an art form, eventually you will learn the meanings of its different patterns, movements or sounds. You will understand what the performers are trying to express. Classical music may sound boring until you study its history and learn to pick out what the different instruments are doing.

Great works of art can express strong emotions and have great beauty; some examine human relationships in a creative way. Many people think these works of art are the finest achievements of the human mind. A play by Shakespeare, a Beethoven symphony or a sculpture by Michelangelo are artistic creations which have become part of our culture and survived through the centuries. The ideas which these works of art express do not seem to date. They lie waiting to be rediscovered by each new generation.

▷ Some pavement artists make a living by reproducing famous paintings with chalk. We appreciate both their skill and seeing a familiar image in everyday surroundings.

PRASAD PRODUCTIONS PVT. LTD.
present

Udhar KA Sindur

PRODUCED BY
L.V.PRASAD

DIRECTED BY
CHANDER VOHRA

MUSIC
RAJESH ROSHAN

◁ When people saw the first movies 100 years ago they could not believe their eyes. They looked behind the screen to see what was there. In the 20th century the cinema has developed as a powerful new medium of entertainment and artistic expression.

a special quality that makes it stand out and last. Charlie Chaplin, Walt Disney's *Snow White* and the Beatles will probably be popular for generations to come.

One culture or many?

One culture sometimes seems to be dominating all the other cultures in the world. This is the culture created by the entertainment, fast-food and fashion industries of the United States of America. These industries are so successful that American film stars, and products such as Coca-Cola, McDonalds hamburgers and Levi jeans, are known and in demand in almost every country on Earth.

Yet at the same time as some young people in jeans queue up for their burgers and fries all over the world, others are striving to keep their local cultures in the face of pressures from big business.

Popular culture

A distinction is often made between the high-brow culture of works of art and the popular culture of everyday life. Popular culture includes the pop songs, TV soap operas, films, computer games and fashions that feature in our daily lives. Most of these things are 'here today and gone tomorrow', like the latest gossip at school. They are fascinating at the moment, but have no lasting value for the human species. In 10 or 20 years' time most of the fashions, programmes and musicians that we think of as great today will have been forgotten. Some popular culture does survive, however, probably because it has

▷ Cultures are in a constant state of change as new ideas, products and styles spread around the world. There is a danger that the worldwide impact of some cultures, such as the American one, will replace the unique features of other traditional cultures.

Traditional ways of thinking, making music and expressing yourself through art help people to feel that they belong to something special and unique. It would be a shame if this rich variety of human culture disappeared.

The sporting spirit

Human beings are competitive. In times gone by contests such as judo, archery and throwing the javelin were part of the preparation for war. Now, however, we usually take part in sports for their own sake, to keep our bodies fit and to enjoy the excitement of peaceful competition. The urge to compete is reflected in our love of sports of all kinds, from simple running races to ball games with complicated rules, such as cricket, that may last for several days at a time.

Sport is now a significant part of culture in the modern world. Supporting your local basketball team or football team is a talking point at school. At a game you meet up with friends and share in the thrills and disappointments of your team's performance.

As a result of television and advertising, many sporting personalities are now as well known, and as well paid, as movie stars. At the peak of his career in the 1970s the boxer Muhammad Ali was possibly one of the best-known human beings in the world. More people had heard of him than had heard of the President of the United States or the Pope.

Major international sports competitions such as football's World Cup are contested by a small number of elite professional players. But there are other competitions for young people who have worked hard to achieve national standard in their chosen sport. Such competitions give them the opportunity to compete with each other.

Olympic games

The world's greatest sporting event is the Olympic Games, which is held every fourth year. The original Olympic Games were held on the plain of Olympia, beneath Mount Olympus, in ancient Greece. They were great cultural events, with poetry and dance competitions as well as athletic contests. The Greek Olympics were first held in 776 BC and then every four years after that for more than 1000 years, until AD 393.

The Olympic Games were revived as an international sporting event in 1896. The spirit of the modern Olympics is intended to be one of friendly competition between amateur athletes from all nations. Sadly, some of the modern Olympic Games have been spoiled by political differences between competing countries. Today the Games have changed dramatically. Professional athletes can now compete in sports such as athletics and tennis. Television and advertising sometimes seem to dominate the event. But most of the thousands of sportsmen and sportswomen who compete in sports as varied as archery, judo, fencing and rowing are amateurs. Despite the commercial pressures, the Olympics Games and similar events, such as the World Student Games, are opportunities for people from all the different cultures on Earth to meet and to make friendships as they take part in peaceful competition.

▽ A sporting success can make everyone feel good. When Nelson Mandela presented the trophy to the victorious South African rugby team in the 1995 World Cup, it seemed to symbolize the new spirit of unity in the previously divided nation.

INVENTIONS AND DISCOVERIES

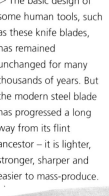

▷ The basic design of some human tools, such as these knife blades, has remained unchanged for many thousands of years. But the modern steel blade has progressed a long way from its flint ancestor – it is lighter, stronger, sharper and easier to mass-produce.

The ingenious human mind has created countless inventions and scientific theories. Many new ideas never work properly and are soon forgotten. Yet every now and then a new discovery, such as the steam engine or the telephone, comes along to change our lives for ever.

What do you think is the most important technological discovery ever made? Is it the axe, clothing, the wheel, the sailing ship, gunpowder, the printing press, steel, electricity, the motor car, the telephone, the computer, the space rocket or the atom bomb? You can probably think of many other possible suggestions. But when our ancestors dicovered how to control the energy of fire half a million years ago (see page 17), they had probably found the key to unlock almost every other discovery.

From people power to solar power

Energy makes everything in the world work. Whenever there is change or movement, one form of energy is being transformed into another. The only sources of energy available to our distant ancestors were the foods they ate and the warmth of the sun. When they wanted to move a load, build a shelter or go on a journey they had to use their own muscle power fuelled by food.

The first new energy source that our ancestors discovered was fire from burning wood. On its own, fire cannot lift loads or drive machines, but it can warm your body, cook your food and light a dark cave. For hundreds of thousands of years after the first use of fire, muscle power

▽ Rapid human progress has been aided by the invention of ways to harness energy sources, other than our own muscles, to do work for us. This time line records some of the major developments in putting energy sources to work.

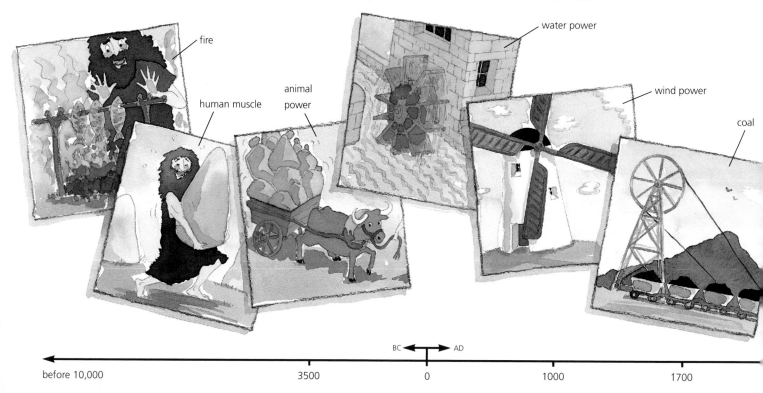

fire

human muscle

animal power

water power

wind power

coal

BC ◀——▶ AD

before 10,000   3500   0   1000   1700

Steam engines use the heat energy released by fire boiling water. The first steam engines were built to pump water from deep tin mines and coal mines. Newcomen's engines were not particularly efficient, but in 1765 a Scottish engineer James Watt developed a much better design. Engines that use the heat from fire to do the work of thousands of people have completely changed the way we live. We use machines that are powered by engines to build our cities, operate our factories, fly us around the world, farm the land, wash our clothes, even to clean our teeth! Today we are more likely to use our muscle power in a game of football or tennis than for harvesting food or building a shelter.

Another great discovery, electricity, brought the power of fuels such as coal, oil and gas (fossil fuels) and, since the 1950s, nuclear fuels, into almost every home. The energy released by burning these fuels in huge power stations is converted into electrical energy, which can be carried along power lines to wherever it is needed.

continued to rule. Then, about 10,000 years ago, when people first started to live as farmers (see page 128), animals were put to work. An ox can haul a plough or turn a water pump, an elephant can lift logs, a horse can carry you on its back. With animal power you can achieve far more than with your bare hands and you may even have time left over from your farm work to do other things.

As civilization developed, people began to make more and more new discoveries. They learned how to use wind and water as sources of energy. The first sailing ships were probably built 5000 years ago by the Egyptians. Two thousand years ago the Romans invented mills powered by flowing water, to grind grain into flour. Windmills were used by the Arabs for the same purpose more than 1000 years ago.

Steaming ahead

When a British blacksmith, Thomas Newcomen, invented the steam engine in 1712, a new energy revolution began.

Today, we have at our disposal many more energy sources than our ancestors did. Electricity powers machines and heats our homes. Oil fuels cars, boats and aircraft. The wind turns mills and pushes sails. The energy generated by power stations has brought great benefits, but

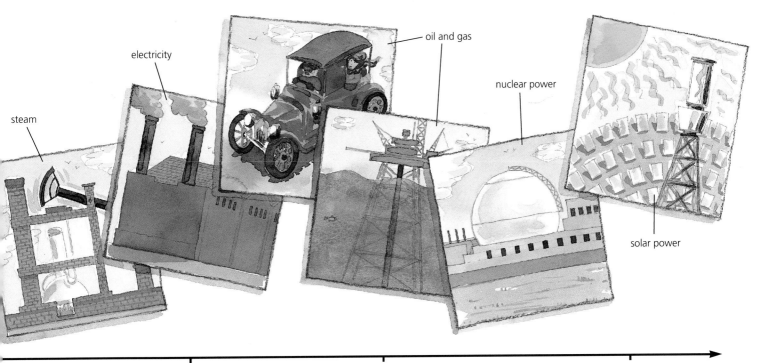

steam   electricity   oil and gas   nuclear power   solar power

1900   1950   1990   2000?

there are drawbacks too. Both fossil fuels and nuclear fuels pollute our environment (see page 152). Also, since none of these fuels is renewable, the world's supply of them will eventually be used up. Scientists are searching for cleaner and more efficient ways to use the almost unlimited energy available in wind, water and sunlight (solar energy). We must find new energy sources to fuel the activities of the growing population of the world.

From donkeys to space shuttles

In the 10,000 years since people first settled down to farm, we have progressed from travelling on our own legs, through riding on carts pulled by animals, sailing with the wind, riding in coal-burning steam trains, to flying in aircraft that are powered by jet engines.

Until the invention of the steam engine, the sailing ship was the most efficient and fastest way of transporting large loads over long distances. Inland, silks and spices were traded across Central Asia for centuries by camels and mules. In remote mountainous areas of the world people still use mules to carry their produce.

The first engine-powered vehicle was an experimental steam train built by a British engineer, Richard Trevithick, in 1804. During the 19th and 20th centuries the growth of railways together with the arrival of the motor car (and motorways) have made it possible for almost everyone to travel long distances over land both quickly and easily.

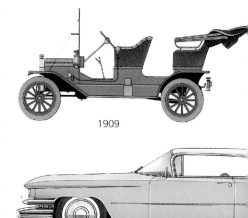

1909

1959

But perhaps the greatest transport revolution of all began at Kitty Hawk, North Carolina, USA on 17 December 1903, when the Wright Brothers made the first powered flight in an aircraft. Suddenly human beings could fly! In no more than the lifetime of one person, air transport has developed from a new invention to a mass transport system which carries millions of people around the world every year. Air transport has made it possible for you to make a journey in a few hours that would have taken your ancestors many months.

From flints and sticks to steel and plastic

If human beings from the prehistoric world came into your house, they would not recognize most of the objects there. They would also not recognize the amazing materials from which the objects are made. Our ancestors used natural materials such as stone, wood, grasses and skins to make their tools and clothing, and to build their shelters. The materials such as metals, pottery, glass and plastics that we use today have been manufactured.

The key to manufacturing new materials was the discovery that natural materials can be changed by heat. When clay is baked in a hot oven, for example, it changes from a soft material that can be shaped, into a hard waterproof material that makes good pots for storage and cooking. Clay pot-making was the first industry to use a manufactured material.

If wood is burned slowly under a mound of soil it changes into charcoal.

INVENTING THE WHEEL

The invention of the wheel is often described as the greatest turning-point in the history of land transport. On your own, you can move comfortably at a rate of about 6 kilometres an hour and carry a load of perhaps a quarter of your own body weight. With the aid of wheels, you can transport much bigger loads on a cart or move your own body much more rapidly, for example on a bicycle. We know that the Sumerians who lived in Mesopotamia 5000 years ago (see page 129) made wheeled carts. They were already using animals to haul ploughs, and these were no doubt soon harnessed to the wheeled carts to transport loads.

▷ Travelling on wheels is far more efficient than using your legs alone. This cyclist can travel about four times faster under his own power on his high-technology bicycle, than he could by running.

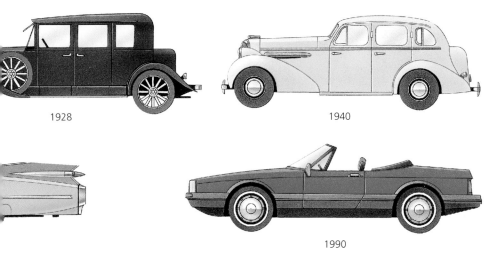

1928

1940

1990

The power of science

Today, developments in technology and science seem to go hand in hand. We take it for granted that a new scientific discovery will probably lead to inventions that will affect our lives. Our scientific knowledge means that we no longer have to rely on trial and error to create inventions. By a combination of careful experiments and creative thinking we have invented computers, released the energy locked in the atom and even discovered the genetic code (see page 13) on which all life is based. This knowledge gives us tremendous power over our lives and over the rest of the living world. With this power comes a responsibility to use it wisely.

When charcoal is burned with certain rocks called 'ores' it releases metals from these rocks. This process is called 'smelting'. The first metal to be widely used was bronze, which is a mixture of copper and tin. (The Bronze Age began about 5500 years ago.) Iron is stronger and harder than bronze, but a higher temperature is needed to smelt it. People first discovered how to smelt iron about 3500 years ago.

Today we use heat energy to produce hundreds of new materials. We have discovered many more metals, some with amazing properties such as the lightweight alloys (mixtures of different metals) used to build aircraft. We can change oil into plastics, drugs and synthetic fabrics. We can even make materials that have the same properties as human bones, and use them to make artificial joints to repair damaged ones.

△ The history of transport is that of a constant quest for speed, efficiency, economy and safety. This quest is reflected in the evolving design of the modern motor car.

▽ New materials lead to new designs. This bridge at Ironbridge in Shropshire, England, was the first bridge in the world made from iron. It was built in 1779 at the start of the Industrial Revolution.

▽ With the aid of wonderful inventions, such as telescopes and microscopes, humans have discovered the structure of the Universe and the building-blocks of life.

EXPLORATION AND ADVENTURE

Do you long for excitement and adventure, or would you prefer to stay safely at home? The human urge to explore unknown places and face new challenges is as strong today as it has ever been.

Long ago, human beings settled in all the comfortable places on Earth – places where there is a good climate, plenty of food and fresh water. But as human populations have expanded, so people have been driven to move to more inhospitable and challenging places – they have set out on incredible sea voyages or trekked across wild and rugged landscapes looking for new places to call 'home'. Today, we live in most of the habitable land areas on Earth, from the desert sands of Africa and Asia to the Arctic wastes of Siberia and northern Canada. But there are some places where the conditions are just too extreme for humans to have settled – impenetrable forests in South America, high mountain peaks buffeted by continuous gales and the icy wastes of Antarctica. These are the wildernesses that attract adventurers like magnets. They hold undiscovered secrets and challenge human beings to test themselves to their limits.

The land of ice

Only one of the Earth's seven continents is so inhospitable that no permanent population lives there – Antarctica. Today, about 2000 scientists live and work in Antarctica during the summer in experimental research stations. Just a few of these people stay through the winter, when temperatures can drop below -60°C. The first human being to be born on the continent of Antarctica was Emilio Palma on 7 January 1978, at Argentina's Esperanza base.

Antarctica was not discovered until the 1820s. The first people to see the continent were probably Nathaniel Palmer, the captain of an American sealing ship, and two British naval officers, William Smith and Edward Branfield. On 7 February 1821 Captain John Davis, another American sealer, made the first known landing on the continent. In the following 180 years of exploration some of the great human

△ In 1914 the Irish explorer Sir Ernest Shackleton led an expedition to make the first crossing of Antarctica. His ship, HMS *Endurance*, became trapped in the ice and was eventually abandoned. Shackleton and his men made an epic journey in small lifeboats through the ice floes to Elephant Island. They were finally rescued in August 1916.

adventures of modern times have taken place in Antarctica.

In 1910 two expeditions set out on what turned into a race to be the first to reach the South Pole. One was led by Norwegian explorer Roald Amundsen, the other by British explorer Captain Robert F. Scott. Amundsen's expedition used teams of dogs to haul their sleds. They were the first to reach the South Pole, on 14 December 1911. They returned safely to their ship. The members of Scott's expedition hauled their loads by themselves. They reached the Pole one month after Amundsen, but on the way back they ran out of food and fuel. After a terrible struggle, they all died.

The scientists who now inhabit and study Antarctica receive their supplies by ship and plane, and move around on motorized snow buggies. At the permanent scientific base at the South Pole, they live in comfort inside centrally heated huts buried under metres of snow. At these low temperatures hypothermia (see page 63) and frostbite (frozen body tissue) can take hold in minutes if you are not properly prepared, so when the scientists are outdoors, they are well protected by modern thermal fabrics which help to retain body heat.

The adventure continues

Antarctica still attracts adventurers. In 1992 two Britons, Ranulph Fiennes and Mike Stroud, set out to walk across Antarctica, pulling everything they needed for the journey behind them on sleds. They suffered incredible hardships, but after walking for 95 days in temperatures below -40°C they had crossed from one side of the continent to the other, via the South Pole. At the end they were close to starvation but were able to radio for an aircraft (something that Scott was unable to do in 1910) and were picked up from the frozen ice-shelf.

Why did they do it? Human beings have a strong urge to compete both with themselves and against Nature. Your personal challenge may be to act in a film one day, go hang-gliding, start your own business or become a successful sportsperson. But for some people these are 'ordinary' challenges and they seek something more thrilling. In the past, these were probably the people who set out to discover new

△ Ranulph Fiennes hauling a heavily laden sled on his 1992 expedition with Mike Stroud to walk across Antarctica.

lands. Today these adventurers pit themselves against dangerous wildernesses, trying to achieve something that no human being has ever done before.

Conquering Everest

There is not much evidence of people climbing mountains for the sake of it before the 18th century. Life then was probably quite challenging enough! But as the way of life became more settled and comfortable in Europe, many people began to travel to different places for the first time. Among the popular destinations were the Alps in Switzerland and Austria, where mountaineering began. Climbing the ice-capped alpine peaks with ropes and ice-axes developed into a popular and challenging sport.

The ultimate challenge to the mountaineer, of course, is to climb the highest mountain in the world. On 29 May 1953, Edmund Hillary from New Zealand and Tenzing Norgay from Nepal were the first people to set foot on the summit of Chomolungma, or Mount Everest, in the Himalayas – the world's highest peak.

▷ There are still discoveries to be made below ground. Cavers explore pot-holes and underground rivers, hoping to be the first to enter a new cavern or find a new route.

They carried oxygen to help them breathe in the thin atmosphere, and were supported by a large expedition of mountaineers and porters as they climbed. Since then Everest has been climbed many times. As mountaineering equipment and our knowledge of the human body's capabilities have improved, so the way in which Everest is tackled has changed. Smaller groups carrying lighter equipment attempt more and more difficult routes to the summit. In 1978 Reinhold Messner from Italy and Peter Habeler from Austria climbed Everest without using oxygen cylinders. Then, in 1980, Messner became the first person to climb Everest alone.

To this day, when asked why they put themselves in such danger to climb to the top of a mountain, the best answer that mountaineers can give still seems to be 'because it's there'.

Mysteries of the deep

Nearly three-quarters of the Earth's surface is covered with water. For thousands of years we have sailed our boats across the oceans and lowered our nets into them to catch fish. Yet human beings have evolved as land animals, and the oceans remain dangerous, mysterious places where we need to use all our intelligence to survive.

In the Middle Ages people generally believed that the world was flat (although nearly 2000 years earlier the ancient Greeks had known that the earth was a sphere). They thought that if you sailed far enough then eventually you would reach the edge of the world and fall off! This belief was finally disproved by the great voyages of discovery made during the 15th and 16th centuries. On an expedition to search for new trading routes to the East

the Italian Christopher Columbus, who did believe that the world was a globe, sailed westwards. He discovered the Bahamas in 1492, although at first he thought he had arrived in India. On a later expedition he became the first European to set foot on the continent of South America. In 1497 another Italian, John Cabot, was the first European to reach North America since the Vikings had briefly set up colonies there 500 years earlier. Cabot thought he had arrived in north-east Asia!

Between 1519 and 1521, an expedition led by the Portuguese explorer, Ferdinand Magellan, made the first circumnavigation of the world, setting out towards the west and returning from the east. Only 35 of the 268 men who set out on the voyage

▽ A single-handed yacht voyage around the world is a tremendous personal challenge.

survived. Magellan himself was killed in a fight with local people when his expedition reached the Philippines.

Scientists sometimes seem to know more about the distant stars in space than about the world beneath the ocean waves. Modern technology is changing this. Sonar sounding instruments and deep-sea submersibles equipped with lights and cameras are exploring the ocean depths. They reveal that the ocean floors are not flat but an incredible landscape of mountains, deep trenches and erupting volcanoes.

Underwater cities

Is there a chance that human beings might one day return to living in the sea? (Animals began to move from water to land about 400 million years ago.) It is unlikely that we could survive on the deep ocean floor. Animals that have evolved in such deep waters have bodies which are adapted to the fantastically high water pressure down there. Our delicate human bodies would be crushed to pulp by the pressure of so much water above us. If we are ever going to colonize the deep oceans we shall have to live in specially constructed cities with immensely strong walls to withstand the pressure. It might, however, be possible to build underwater cities in the shallow seas bordering the continents. Some people have even imagined that, in the future, scientists might be able to alter our genes to make us capable of breathing through gills.

To the stars

When the spacecraft Apollo 8 left the Earth and travelled towards the Moon in 1968, it was the first time that human beings could look down from space and see their home planet, the Earth. Many people regard this moment as a turning-point in human history. Now that we have all seen photographs of the planet Earth from space, we realize that it is not such a huge place after all. The world is getting more crowded and more affected, even harmed, by human activities all the time. At the end of the 20th century we are looking at space as a possible future home for humans.

During the 1990s, the pace of space exploration appears slower. But space

▷ Between 1969 and 1972 six Apollo missions took men to the Moon. The 12 astronauts who landed there are the only human beings to have set foot on another world. Within a few generations many more men and women will probably have set out on journeys to the Moon, Mars and beyond.

technology is developing steadily. Unmanned spacecraft such as the Viking landers and the Voyager probes have visited the planets of the Solar System, sending back incredible images and streams of scientific data. People have continued to travel into space aboard space shuttles and the Soyuz spacecraft. The Americans and Russians are planning and developing permanently occupied space stations to orbit the Earth. When humans do return to the Moon, probably some time early in the 21st century, they will set up a permanent scientific base. Not long after, a journey will probably be made to set up a colony on Mars, and for the first time human beings will be living on another planet. In a few centuries' time, the gap between the first journeys to the Moon and the first human settlements in space will not seem so long. After all, more than 100 years passed between Cabot's voyage to North America and the arrival of the Pilgrim Fathers from England to settle there in 1620. It is more than 5000 years since the first civilizations arose on Earth (see page 129). It is just 30 years since human beings first travelled into space. In the history of the world these are incredibly short times. Who knows where our urge to explore will have taken us in another 5000 years' time?

THE FUTURE

What does the future hold for human beings? Will the human branch of the tree of life grow straight and strong? Will it divide, or perhaps even wither and die?

Has evolution stopped with the human being? The thought behind this question is easily understood. When you read about the history of life on Earth, it can seem like a story of steady progress towards the human species. At first there were only simple, single-celled living things which later became more complex. They developed and branched into the five kingdoms of life (see page 9). Later still, fish developed into amphibians which emerged from the water on to land, dinosaurs rose and fell, and mammals appeared on Earth. Finally, at the top of the evolutionary tree, appeared the human being. It can seem as if evolution has now reached its goal.

This is a very mistaken view. There is no evidence that evolution has a particular direction or goal and there is no reason to suspect that it stops with human beings. One possible cause of this misunderstanding is that human evolution takes place over much longer time-spans than individual human lives. During your lifetime, or even during 100 successive human lifetimes, there will not be enough time for evolution to produce dramatic changes to the human species. It can look as if nothing much is happening on the evolutionary front.

The tree of life

To give you an idea of the time-scales involved in evolution, imagine the tree of life (see page 11) as an oak tree 50 m tall. This tree has taken 3500 million years to grow from the ground (the first living things) to its present height. The branches and twigs that represent all the different dinosaur species that ever lived did not start to grow until the tree was just over 47 m tall. For 150 million years they grew strongly and then, when the tree was just over 49 m tall, this part of the tree died. This represents the time when the dinosaurs became extinct. A nearby branch then started to strengthen and divide – this is the part of the tree that represents mammals. It is still growing strongly today. Right at the end of the mammal branch is a twig just 1.5 cm long. This represents the human genus *Homo*.

Today there is just one living bud left on the *Homo* twig. It is now just 2 mm long. This tiny bud on the tree of life represents the whole history of the modern human being, *Homo sapiens sapiens*. Your whole life is represented by just one microscopic cell one-millionth of a metre thick on the very tip of this bud. It is hardly surprising that evolution seems so slow from a single human point of view. Who knows what will have become of our species when the tree has grown another metre, or even just a few more millimetres?

▽ A busy street in India. On the time-scale of evolution the growth of cities has been like a sudden explosion. What does the future hold for human beings?

Is the human being special?

Evolution is a random process, like drawing the winning numbers in a lottery. If you could go back in time and rerun evolution, there is no guarantee that the world would end up with anything like the same species that have evolved this time around. Certainly, an intelligent species capable of technology, language and art might evolve, but would it be anything like us?

Of course we think we are special because we are us. Just as your own family is more special to you than other families, humans as a whole are more special to us than other species. We also think we are special because we seem to be so intelligent and adaptable. Unlike every other species, we can use our intelligence to alter our environment dramatically and to plan our future.

On the other hand our intelligence could be our downfall. It could be an evolutionary deadend. Our intelligence has allowed us to expand and develop to the point where our activities are changing the world we live in. As the human population grows, our industries and activities threaten the environment that keeps us alive. We have invented weapons which we could use to destroy ourselves, making way for less intelligent but more successful creatures to live on Earth. We probably stand a good chance of avoiding complete destruction of ourselves. But we may still have some terrible lessons to learn before we discover how to take proper care of our environment and keep our aggression under control.

Facing the future

We can only really dream and speculate about our long-term future. It would be wonderful if humans did colonize space, cure cancer, solve the problem of famine and learn to live in peace. We will take steps towards some of these goals in the 21st century, and it will be incredibly exciting to see the progress that is made during your lifetime.

In the short term, however, human beings have to face some urgent problems. The one problem that lies at the heart of many others is the rapidly expanding human population.

Population explosion

The human population is growing faster than ever before. In 1990 the population of the world was estimated to be 5.3 billion. By the year 2000 it will have reached 6.2 billion and by 2025 it could have reached 8.5 billion. During your lifetime the world's population will more than double.

When a population grows, it means that more people are born and survive each year than die. Most animal populations change from one year to another: in a good year more animals survive and the population goes up; in a bad year more die and the numbers fall. If all the ducklings (or even most of them) that hatch each year were to survive, then after a relatively few years the world would be knee-deep in ducks. The duck population is kept in

△ This kind of pollution can effect the health of people living nearby. As planet Earth becomes more crowded each year, we must take increasing care to protect the environment in which we live.

check because most ducklings are killed in the first few weeks of life by predators, bad weather or a shortage of food.

Sometimes animals find themselves in an environment in which there are few natural predators. For example, when European settlers moved to Australia they took rabbits with them. No foxes or other natural predators of rabbits lived there and so the rabbit population exploded. Such natural population explosions end, however, as food supplies run out or a new disease attacks the population.

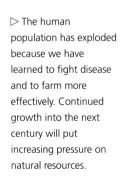

▷ The human population has exploded because we have learned to fight disease and to farm more effectively. Continued growth into the next century will put increasing pressure on natural resources.

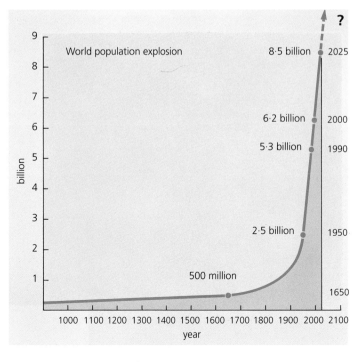

World population explosion

8·5 billion — 2025
6·2 billion — 2000
5·3 billion — 1990
2·5 billion — 1950
500 million
1650

billion

year
1000 1100 1200 1300 1400 1500 1600 1700 1800 1900 2000 2100

Crowded planet

Humans have few natural predators. Our hunter-gatherer ancestors lived in small groups because food supplies were restricted. For most of human history the population may have been no more than a few tens of thousands. Then, about 10,000 years ago, agriculture was discovered. This was the start of the first human population explosion: the size of the world's population began to rise. Sometimes it fell as a result of diseases such as the Black Death, but the trend was upwards. By the 17th century there were about 500 million people in the world (about the same number as there are in Europe today). A second, more dramatic population explosion was triggered by the improvements in technology, industry and scientific understanding that started in the 17th century. These led, in turn, to the great migrations from Europe to the New World. Between 1650 and 1950 the world's population increased from 500 million to 2.5 billion. Since the 1950s it has been growing even faster. This is because modern medicines, new crop varieties and pesticides have improved the health and nutrition of people throughout the world.

Freedom of choice

Most of the current population growth is taking place in the developing countries, where people are poorer. The population of Africa, for example, will probably triple between 1990 and 2015. In the developed countries, populations are growing only very slowly or in some cases even falling, as in Sweden. You may find it surprising that poorer people have more children than wealthier people. Wealth allows people to choose. In developed countries people have access to the education and medical advice that allows them to choose how many children to have. One of the great problems for the future is to make this knowledge and support available to people in the developing world.

At present agriculture can provide enough food to feed everyone in the world, but there is a problem of distribution – getting the food to all the people who need it. There is famine in one place while some people have surplus food in another. But, even if distribution problems

POISONING OUR WORLD

Population growth is also a worry because of its effect on the environment. In the developed countries we are just starting to understand the potentially disastrous effects that our own activities are having on the world as a whole. The gas carbon dioxide occurs naturally in the atmosphere and helps to keep the Earth warm by trapping heat from the Sun. Human use of fossil fuels, such as coal, gas and oil, in cars, factories and power stations in particular, is now increasing the amount of carbon dioxide in the atmosphere above natural levels. This is called the 'greenhouse effect'. This traps extra heat, leading to a general increase in the Earth's temperature. As a result of this 'global warming', the Earth's climate may change, fertile land may be transformed into desert, and islands and coastal areas may be flooded as sea levels rise. The scale of the problem is likely to increase as the populations of the developing countries grow. As their standards of living improve, so each person uses more energy as they heat and light their home, buy more consumer goods and drive a car.

△ Our use of coal, gas and oil is changing the Earth's climate. Finding clean alternative energy sources, such as these windmills, must be a priority in the 21st century.

can be solved, it will not be too many years before the continued population growth leads to widespread food shortages once again.

Conflict

Some history books seem to be a never-ending catalogue of wars, conquests and human conflict. The past 100 years have seen some of the most destructive wars in history. At the end of the 1990s there are fewer conflicts around the world than for some time. Yet the stockpiles and destructive power of weapons are greater than ever before. At present the threat of nuclear war seems less strong than it did 20 or 30 years ago. But now that nuclear weapons have been invented they cannot be 'un'-invented. Who knows what conflicts will arise in future centuries? Let us hope that we can use our human intelligence to prevent conflict from destroying the entire human species.

One family

These last few paragraphs have discussed some of the problems that human beings must face if we are to make the most of our potential as a species. There are many positive things to report too. During the same century that human minds have created weapons of mass destruction, they have decoded the thread of life (DNA, see pages 12–13) which shows that all human beings are part of the same family. We have invented drugs to cure terrible diseases, produced new crops to relieve famine and built spacecraft to explore the universe beyond the Earth. The communications revolution is carrying this knowledge all around the world. For the first time we are all aware of each others' problems and are concerned to find solutions to them. Let us hope that we can use our incredible minds to create a peaceful human family to which we are all proud to belong.

▷ It is always better to talk than to fight. Perhaps computers and the worldwide exchange of information will change the world for the better.

▷ Every person is a wonderful unique individual, yet we are all members of the one human family. We share our ancestors, and the inheritance they left us in our genes, with every other human being on Earth.

GLOSSARY

AIDS Acquired Immune Deficiency Syndrome. A disease caused by the HIV virus, which is transmitted from one human being to another by contact with infected blood and other body fluids.

amino acids Molecules which link together into long chains to build all the various protein molecules in the human body and in other living things.

antibodies Substances produced by the body's immune system that help to recognize and destroy harmful bacteria and viruses.

antibiotic A drug which kills bacteria and other organisms that infect the body.

artificial intelligence The attempt to reproduce the intelligence of living things, especially human beings, with electronic computers.

Big Bang The theory that our Universe was created by an incredible explosion of energy and matter about 15 million years ago and that it has been expanding and cooling ever since.

carbohydrate Sugars and starches which the human body uses as fuel.

carbon dioxide A gas composed of carbon and oxygen which human beings, and other living things, produce as a by-product of respiration.

cell The basic microscopic living unit from which living things are built.

chromosome A tightly coiled strand of DNA. It carries the genes which determine all our inherited characteristics. Human cells contain 46 chromosomes.

circulation The system consisting of the blood vessels, blood and heart, which transports cells, raw materials and wastes around the human body.

civilization The settled and organized way in which large human groups live together with agreed leaders, laws, customs and traditions.

culture The beliefs, traditions, art, customs and activities which the members of a human group share.

dehydration Excessive loss of water from the body.

digestion The process by which the body breaks down food into the raw materials it uses to build and repair its tissues, and to provide the energy to maintain its activities.

DNA Deoxyribose nucleic acid. The long spiral strands which carry the genetic information that builds the human body and other living things.

element A substance consisting of just one type of atom. There are about 100 different elements, and all material objects on Earth, including living things, are made from them.

energy The capacity which some things have to do work or bring about change. Food provides the energy to maintain the activity of the human body. Fuels such as coal and oil provide the energy to power machines.

enzyme A particular type of chemical which speeds up life processes in the body, for example digestion.

evolution The process by which living things have changed and new species have emerged.

gene A section of the genetic information, carried by the chromosomes in cells, that determines a difference between individuals. For example, one gene may determine your eye colour. You inherit your genes from your parents.

genetic clock The number of differences between the DNA of two individuals or species used as a clock to reveal how long they have been evolving from a common ancestor.

genetic code The simple code in which the genetic information carried by DNA is written. The code is formed by the order of chemical groups called bases which hold the two spiral strands of DNA together.

genus A grouping of closely related species all descended from a common ancestor. Modern humans are the only living members of the genus *Homo*.

gland An organ which manufactures a substance that the body needs. For example, salivary glands in the mouth and throat produce saliva.

HIV The Human Immunodeficiency Virus. The virus that causes the disease AIDS.

homeostasis The constant internal condition that the body maintains by controlling its own temperature and chemical composition.

Homo The human genus.

Homo erectus A human species that originated in Africa about 1.5 million years ago and spread into Europe and Asia before dying out about 0.5 million years ago.

Homo habilis The first known human species. Homo habilis lived in Africa from about 2.5 to 1.5 million years ago.

Homo sapiens neanderthalensis A species of human being that spread through Europe about 130,000 years ago.

Homo sapiens sapiens The modern human species.

hormone A chemical produced by the body which acts as a messenger by stimulating the activities of certain cells.

hunter-gatherer The human way of life for most of human history, in which food was obtained by hunting wild animals and gathering wild plants.

hyperthermia Increased body temperature above normal.

hypothermia Cooling of the body below normal body temperature.

immune system The body's internal defences against infection. Invading microbes are recognized and attacked by a variety of different white blood cells.

infection Invasion of the body by foreign organisms.

innate Describing an ability or skill that you are born with and do not have to learn.

Internet The global computer network, based on the telephone system, which individuals can use to exchange messages and search for information.

kilocalorie A measure of the energy content of food.

lipid A molecule of fat.

lymphocytes White blood cells which help the body recognize and fight infections by foreign organisms.

meiosis The special type of cell division that takes place in the ovaries and testes, to produce sex cells with half the normal number of chromosomes.

membrane A thin skin or wall surrounding a cell, part of a cell or some other structure in the body.

microbe Any microscopic organism, for example bacteria.

minerals Simple inorganic (non carbon-based) substances, such as common salt and iron, that your body needs to stay healthy.

mitosis The normal process of cell division in which a single cell divides into two copies of itself, each with a full set of chromosomes.

molecule A group of atoms linked together by chemical bonds.

monera The kingdom of microscopic living things that includes bacteria.

multicellular Describing an organism made from many different cells that work together as a single living thing.

natural selection The process by which individuals that are better adapted or more fitted to their environment are more likely to survive and to reproduce.

nervous system The brain, spinal cord, peripheral nerves and sense organs that together form the body's main sensing, control and decision-making system.

neurotransmitter One of a number of chemicals that carry nerve signals from one nerve cell to another.

non-verbal communication Communicating without words, for example with body language.

organ A specific part of the body with one or more jobs to do, for example the heart or the liver.

organism A single living thing.

perception The process in which the mind makes sense of signals from the sense organs.

phagocytes White blood cells that gobble up foreign invaders in the body.

phoneme One of the the basic sounds we use in speech.

protein A complex molecule built from chains of amino acids. The body manufactures thousands of different proteins which perform many different jobs. Proteins make muscles contract, give body tissues strength and structure, and transport substances around the body and in and out of cells.

protist The kingdom of microscopic living things that includes amoeba.

psychoanalysis A method for treating mental problems. With the help of a therapist the subject tries to understand his or her 'unconscious' thoughts and feelings.

puberty The stage of life during which our bodies become sexually mature.

reflex actions Automatic actions, for example blinking in response to a bright light, that take place without conscious thought.

religion Human beliefs about a god or gods and the meaning these beliefs give to human existence. Members of a particular religion share the same beliefs and express them through rituals.

respiration The chemical process by which living things release energy from their food. During normal human respiration, oxygen absorbed into the blood as we breathe is combined with the sugar glucose.

ritual A ceremony or practice performed as part of religious worship or to mark some special occasion.

RNA Ribonucleic acid. A molecule, closely related to DNA, that carries the genetic information in some viruses. RNA also plays an important part in the construction of proteins from the genetic plan stored on the DNA in human cells.

selective breeding The process by which human beings select which plants or animals to mate in successive generations to produce offspring with certain characteristics, for example fast horses or highly productive crops.

sense organ An organ such as the eye or the ear that detects information about the environment outside the body.

sexual reproduction Reproduction in which a new individual is produced when sex cells from a male and a female combine.

species A group of similar living things which are capable of breeding with each other, but not with members of other species.

subsistence farming Farming in which almost all the food produced is needed for the farmer's family to survive.

tissue A group of linked cells in the body that performs a particular task, for example muscle tissue.

unconscious Describing thoughts and feelings that are hidden or suppressed from our consciousness.

vitamin Essential substances that we require in small quantities for health.

INDEX

Where several page references are given for a particular headword, the more important ones are printed in bold (e.g. **41**). Page numbers in italic (e.g. *94*) refer to illustrations and captions.

ACKNOWLEDGEMENTS

Design: Keith Shaw
Picture research: Penni Bickle
Abbreviations: t = top; b = bottom; l = left; r = right; c = centre

Photographs
The Publishers would like to thank the following for permission to reproduce the following photographs:

Heather Angel/Biofotos: 10tr; 15b
Allsport: 141b (Francois Pienaar); 144b (Bruno Bade/Vandystadt)
Agence Vandystadt: 148 (Thierry Martinez)
Amnesty International: 135t
Ancient Art and Architecture Collection: 22t (B Wilson); 118t; 120; 121b (Ronald Sheridan)
Anthony Blake: 36b (Merehurst)
Archiv der Gesellschaft der Musikfreunde in Wien: 107t
British Heart Foundation: 47r
The Bridgeman Art Library: 92b (Archimboldo); 94t (Roy Miles Gallery)
Julian Cotton Photo Library: 15t; 32t; 36t
Cordon Art: 97 (1996 MC Escher/Cordon Art–Baarn–Holland. All rights reserved)
James Davis Travel Photography: 128t, 129
Environmental Picture Library: 150t; 152 (Daniel Beltra)
Mary Evans Picture Library: 41b; 82; 107b (JL Charmet)
Eye Ubiquitous: 10 (Keith Mullineaux); 37t (Thelma Sanders); 41tr (Paul Seheult); 51b (Sue Passmore); 70t (1994 Vidal) b (Roger Chester); 71m (Paul Seheult) b (Tim Page); 72 (Stewart Weir); 73 t; 77b (David Cumming); 92r (Darren Maybury); 99 (Matt Wilson); 101 (Mike Southern); 103b (Dorothy Burrows); 108t 109t (Bennett Dean); 126b (Matthew McKee); 133b (Julia Waterlow); 136b (R Donaldson)

The Jane Goodhall Institute: 13 (Hugo van Lawick)
Sally and Richard Greenhill: 112t
Ronald Grant Archive: 104; 136t; 140t
Michael Holford Photography: 142t
Hulton Deutsch Collection: 123; 146b
The Image Bank: 6-7 (Romilly Lockyer); 9t; 26-7 (J Freis); 29 (Janeart Ltd); 53 (Al Tielemans); 54; 66b (Marc Romanelli); 69 (Alvis Upitis); 81l (Paul Trummer); 83b (Gary Cralle)
Image Select: 6r (Allsport/Mike Powell; 20t (Ann Ronan/Jacana/Herv Chaumeton); 23 (Ann Ronan/Jacana/Jean-Michel LABAT); 62t; 63b (Allsport/Mike Powell)
Impact Photos: 2 (Jez Coulson); 4 (Mark Henley); 5t (Jez Coulson); 57t (F Perri); 59b (Martin Black); 95b (Bruce Stephens); 109b (Jez Coulson); 116-7 (Homer Sykes); 119b (Gideon Mendel); 127b (Javed A Jafferji); 131t (Mark Henley); 140b (Mark Henley)
Magnum Photos: 1 (Burt Glinn); 113t (Michael Nichols); 134 (B Barbey); 139t (Burt Glinn); 147b (Michael Nichols)
Material World: 122b
Oxford Scientific Films: 9m (Alistair MacEwan)
Performing Arts Library: 48b (Clive Barda)
Pictor International: 112
Range Pictures Ltd: 125t (Corbis-Bettmann)
Red Cross: 75
Redferns: 55 (Bill Davila)
Rex Features: 21T (Sipa/Patrick David); 33b (Philip Reeve); 145bl &r
Chris Ridgers: 94b
Royal Geographical Society: 147t (Ranulph Fiennes)
Phil Schermeister: 97b
Science Photo Library: 9b (A B Dowsett); 12t &b (Dr Jeremy Burgess); 14 (John Reader); 16t (John Reader) b (Martin Dohrn); 18; 22b (John Reader) 28t (Dick

Luria) b (Nasa); 35t (Jerry Wachter) b (Dick Luria); 39b (Ron Sutherland); 41l (Biofoto Associates); 42 (Robert Becker/Custom Medical Stock Photos); 43l&r; 44 (Prof P Motta/La Sapienza University/Rome) b (Jerry Mason); 46 (Dept of Clinical Radiology, Salisbury District Hospital); 47l (Chris Priest & Mark Clarke); 50 (Prof C Ferland); 52 (Michael Burgess); 56b (Francis Leroy); 57bl (John Barosi) br (Prof P Motta); 60 (Mark Clarke); 63t (Dr R Clark & MR Goff); 67b (Richard Rawlins/Custom Medical Stock Photos); 68 (CNRI); 74 (James King-Holmes); 76 (CDC); 80t (Paul Shambroom) b (NIH/Custom Medical Stock Photos); 81r (JC Revy); 84-85 (Scott Camazine); 86m&b (CNRI); 89t (Wellcome Dept of Cognitive Neurology); 90 (Peter Menzel); 91 (Dr Jeremy Burgess); 106 (J L Charmet); 121t (Philippe Plailly/Eurelios)146t (Nasa/SPL); 153b (J Okwesa)
Frank Spooner: 33 (Frank Spooner/Gamma); 59t (Hemsey-Liaison)
Still Pictures: 8t (Mark Cawardine); 10b (H Ausloos); 64 (Mark Edwards); 78 (Jorgen Schytte); 125b (Adrian Abib); 126t (Carlos Guarita)
Tony Stone Associates: 2 (Tony Arruza); 73b; 83t (David Madison);110t (Paul Chesley); 114 (Tony Arruza)
Telegraph Colour Library: 89br (VCL) 102b; 103 (Alan Brooke); 115t (Planet Earth/Andre Bartschi) b (F.R.G. F McMckinney); 131b (Dilip Mehta/Contact);132 (Roger Garwood); 130 (H Rogers) 132t; 135b (R Searle);

139br (N Mosley); 143 (H Rogers); 149 (Nasa) 150 (H Rogers); 151 (A Kuznetsov); 153 (J Okwesa)
University of Iowa: 89bl (H Damasio)
John Walmsley: 122t; 124tr bl
N Ward: 61r
Stewart Weir: 95t&m; 108b; 119t
Wellcome Centre MedicalPhotographic Library: 100
Zefa Pictures: 5b; 48 (Zefa-BOESCH); 98; 124br (Edgeworth Productions)

Cover
Front, spine and back cover:
The Image Bank: tl (U. Ommer), tr (J. Darell)
Tony Stone Images: bl (Peter Correz), br (T. Madison)

Illustrations and diagrams
Mike Codd: 19
Michael Courtney: 34tr; 40; 45c and b; 55t and b; 68-69b
Chris Forsey: 30-31; 61t; 65t; 77t; 87t; 88
Clive Goodyer: 35bl
John Haslam: 13t
Nick Hawken: 11tr; 17; 94r and b; 95tl; 151
Richard Hook: 21b
Frank Kennard: 34bl; 43b; 53t
Olive Pearson: 24
Mike Saunders: 32; 39tl
Kate Simpson: 11b; 29b; 38t; 62b; 71t; 90b; 99t; 100t; 101cr; 105; 111; 142-143
Steve Weston: 40; 58; 67; 74; 93b
Peter Visscher: 14r

Consultants
The following people read and commented on different chapters in the book: S.R. Milligan, N. Moloney, J. Pope, N. Tucker.